3—

Toby Burrows

The Text
in the Machine
Electronic Texts
in the Humanities

Pre-publication
REVIEWS,
COMMENTARIES,
EVALUATIONS . . .

"Toby Burrows' *The Text in the Machine* is a very timely and well-informed survey of the state of the art in the making and use of electronic texts. It will be particularly useful for people about to begin projects involving electronic scholarly texts in the humanities. Burrows gives a thorough account of what is already a large and fast-moving range of activity. It is clearly written and well documented. This book will serve not only to orient present-day readers new to the field, but will act as a useful record for readers to come of how matters were at an early moment in the history of electronic texts."

Dr. Peter Robinson
Centre for Technology and the Arts,
De Montfort University,
Leicester, England

The Haworth Press
New York • London • Oxford

The Text
in the Machine
*Electronic Texts
in the Humanities*

THE HAWORTH PRESS
New and Recent Titles of Related Interest

The Text
in the Machine
Electronic Texts
in the Humanities

Toby Burrows

The Haworth Press
New York • London • Oxford

The Haworth Press, Inc., 10 Alice Street, Binghamton, NY 13904-1580

Cover design by Jennifer M. Gaska.

Library of Congress Cataloging-in-Publication Data

Burrows, Toby.
    The text in the machine : electronic texts in the humanities / Toby Burrows.
        p.   cm.
    Includes bibliographical references (p.) and index.
    ISBN 0-7890-0424-0 (alk. paper).
    1. Humanities—Electronic publishing.   2. Humanities—Electronic publishing—United States. I. Title.
AZ186.B87 1999
070.5'797—dc21   98-49068
                                                                   CIP

CONTENTS

ABOUT THE AUTHOR

Toby Burrows, PhD, is Principal Librarian of the Scholars' Centre at the University of Western Australia. He has worked with electronic texts in the humanities for the last four years and set up a networked electronic text service for the University in 1996. An editorial board member of *Parergon,* the journal of the Australian and New Zealand Association of Medieval and Modern Studies, Dr. Burrows is also the author of *British University Libraries* and co-editor of *Management in Australia and New Zealand* (both published by The Haworth Press, Inc.). In addition, he is co-director of a nationally funded project that is establishing a World Wide Web service for the Berndt Museum of Anthropology in Perth, Western Australia.

Preface

Most scholarly work in the disciplines collectively known as the humanities is concerned with documents that record the cultural life of the past. Interpreting, analyzing, and responding to the texts contained in these documents is at the heart of scholarship and research in literature, history, philosophy, and similar fields. Traditionally, this work has been founded on a consensus about the nature of texts and their authorship and about the relative significance of different texts and groups of texts. The major questions posed by postmodernist critics have attacked and undermined this consensus:

- What is a text, and what forms can it take?
- What are the respective roles of authors and readers in the creation of texts?
- Is there such a thing as a canonical collection of texts?
- Are some texts privileged over others in their significance and value?

However, the important thing to note is that this debate centers around texts, which assumes their continuing centrality to the humanities. While the nature and significance of texts are debated, their importance is confirmed and reinforced.

The new electronic technologies are affecting profoundly the study and use of texts in the humanities, along with most other areas of scholarship and research. Since the early 1990s, a rapidly increasing number of experiments and projects have been established to explore the implications of electronic formats for texts. Although this activity has mostly been of marginal interest to the humanities for much of that time, it is now moving quickly toward a more central role. Encouraged by the success of major scholarly initiatives such as the *Canterbury Tales Project* and by the emergence of commercial services such as Chadwyck-Healey's Literature Online, scholars are beginning to accept that electronic texts can provide the basis for significant new approaches to their work.

Electronic versions of texts are an extremely useful and important addition to the range of tools available for scholarship in the humanities. This can be said without necessarily endorsing more speculative views about the wider significance of electronic media. My aim here is not to contribute to theoretical polemics and speculations over the future of the book or to discuss the possible replacement of printed and written texts by electronic ones. Nor do I intend to debate the view that computer-based hypertextual composition is a more advanced and powerful method of writing than the familiar linear models. My goal is a practical one: to explain how scholarly electronic texts are constructed and to illustrate some of the scholarly benefits derived from this process.

What is an electronic text? For the purposes of this book, there are two essential characteristics. First, the text must be stored in an electronic format, for manipulation by appropriate software. In addition, this format must also be published or distributed in some way. Private versions of texts, which reside in an individual's machine for personal study and use, are excluded from consideration.

Second, an electronic text must be a version of a text or document that previously existed in print or in manuscript form. Although there are various examples of texts—particularly fictional or poetic—composed in an electronic form, these are outside the scope of this book. The focus here is on the transformation of existing texts into electronic form for the purpose of study and research.

In considering how this transformation can be accomplished, I have tried not to assume too much technical knowledge on the part of my readers. This book is primarily intended for the interested scholar, student, or librarian working in the humanities who wants an overview of the methodologies and issues involved in creating electronic texts. Although I hope that experts in this field may find something of value here, I am aware that, in many areas, I have summarized complex matters—particularly of a technical kind—in a somewhat simplistic way. My primary goals are to guide those who wish to venture into this field and to provide a snapshot of current activity in relation to electronic texts in the humanities.

Several different areas of expertise must be combined to create and distribute electronic texts in the humanities. These are not always found in the same context, but I aim to bring them together here. Electronic texts inevitably have a substantial technical dimension that affects anyone involved with them. Without an understanding of ways of creating and marking up such texts, little progress can be made. These topics are covered in Chapters 1 and 2. Similarly, electronic texts can only be created and used with suitable software, and Chapter 3 contains an introduction to this somewhat labyrinthine subject. For librarians and publishers, important and difficult questions must be addressed in developing an appropriate framework for publishing, collecting, and providing access to electronic texts. Matters of this kind are discussed in Chapter 4. For scholars, one of the most important considerations is the nature of scholarly editing and its relationship to electronic representations of texts; these issues are covered in Chapter 5.

As a whole, then, this book contains an introduction to the processes involved in creating and publishing electronic texts in the humanities. It is intended for all those interested in setting up, managing, or contributing to electronic text projects—scholars, students, librarians, and publishers alike. Such projects are still in their infancy, for the most part, particularly in comparison with their printed predecessors. Yet, they contain enormous promise for the future study and use of texts.

Toby Burrows
The Scholars' Centre
The University of Western Australia

Acknowledgments

My debts in completing this book are many.

My colleagues William Hamilton, Elizabeth Hardy, and Claire McIlroy read draft versions of various chapters and made very helpful comments. Mark Huppert, Matthew Chalwell, and Stephen Trefry gave me technical advice and assistance.

David Seaman provided the initial inspiration and encouragement that started me on this path.

The University of Western Australia provided valuable backing in a variety of ways, including a grant to set up an Electronic Text Service and assistance to attend conferences. Three successive university librarians—Arthur Ellis, Imogen Garner, and John Arfield—have supported my work in this area over the past four years.

Above all, I am very grateful to Tessa and to Emily, Lucy, and Nigel for their continuing love and understanding, despite my, at times, obsessive interest in what seems a highly mysterious subject to them!

Chapter 1

Markup Systems for Electronic Texts

At the most fundamental level, a computer only stores data as binary digits ("bits")—either 0 or 1. Superimposed on this bedrock are various layers of interpretative mechanisms—assembly language, operating system, application program, and the like—that instruct the machine how to combine its bits into a display suitable for the human eye. As far as electronic texts are concerned, the uppermost layer is a representation of the text that can be read on a screen.

In constructing an electronic representation of a text, two basic approaches produce quite different results: representing the text as an image and representing the text as a linguistic entity. The former produces a digital photograph composed of picture elements ("pixels") that represent colors, including black and gray. The latter produces a display that is the equivalent of a printed or typescript text, composed of characters which represent letters and typographical features.

In this chapter, we are concerned with the latter approach. We look first at the types of markup systems that underlie the display of character-based texts and at ways of evaluating and comparing them. The major alternative schemes are then discussed, with particular emphasis on the Standard Generalized Markup Language (SGML) and the Text Encoding Initiative (TEI).

WHAT IS MARKUP?

Markup, in the broadest sense, can be defined as "all the information in a document other than the 'contents' of the document itself, viewed as a stream of characters" (Sperberg-McQueen, 1991:35). In other words, anything other than the actual letters that make up the text should be regarded as markup. Most obviously, perhaps, markup includes all information about the format of the text such as margins, fonts, and page breaks. It also covers all information about the structure of the text such as chapter divisions, paragraphs, headings, footnotes, endnotes, sentences, and even word division.

Markup is independent of the form in which the text is distributed. We are familiar enough with these properties in a printed text, since editors and compositors, as well as authors, have traditionally "marked up" typescripts, manuscripts, or galley proofs to show the format and structure of the final printed version. But markup is present, to a greater or lesser extent, in handwritten texts too. The illuminated manuscripts of the Middle Ages, for example, are full of information about the text, often to such an extent that the text itself becomes almost invisible. This is especially evident in the elaborate decorative initials that, functionally, merely signal the start of a new section of the text. The extraordinarily rich and beautiful decoration used for the words "In principio" at the beginning of St. John's Gospel in the *Lindisfarne Gospels*, for instance, is a marvelous example of the integration of markup and art (Nordenfalk, 1997:73-75).

It is difficult to imagine a text without any markup at all. A possible candidate would be a manuscript that consisted entirely of an unbroken stream of letters, without any word division, sections, or intentional page breaks. Some early Hebrew, Greek, and Latin manuscripts exhibit this approach. However, even here it could be argued that there is still an assumed level of markup embedded in the text, since breaks between words would be interpreted and applied by the reader as part of the process of reading.

For the representation of texts in electronic form, markup is essential. Without it, a computer cannot display a text to a reader in a meaningful way. The type of markup scheme chosen is also crucial, since it determines the eventual appearance of the text and the ways in which it can be used. Sperberg-McQueen (1991:37-44) identifies six fundamental characteristics of texts that should be taken into account in markup schemes:

1. *The linguistic nature of texts.* Since texts are linguistic objects, it should be possible to analyze the linguistic organization of

the text from its markup. Among the features that might need to be discernible are lexical forms of words (headwords in dictionaries), parts of speech, phonetic transcription, and ambiguous meanings.

2. *The physical nature of texts.* The kinds of texts we are concerned with here were originally created and transmitted as physical objects. To represent them in electronic form may require a kind of markup that can record and reproduce the unique physical characteristics of a manuscript or a printed book.

3. *The hierarchical structures of texts.* As well as having a basic linear arrangement, texts have hierarchical structures. In epic poetry, lines exist within verses within books; in tragedies, lines exist within scenes within acts. Much-used and often-quoted texts such as the Bible have a standardized referencing scheme that exists separately from any specific version of the text, and may well conflict with it. A markup scheme must be able to document and reflect the hierarchical structure or structures applying to a particular text.

4. *The internal and intertextual references of texts.* Most texts contain a network of cross-references of some kind, disturbing the linear appearance of the text. These may be references within the text itself, such as links to other sections or endnotes. They may also involve references to other external texts, in the form of quotations, paraphrases, or reproductions. These hypertextual and intertextual features add richness to the text and must be identifiable through markup.

5. *The referential nature of texts.* Texts inevitably refer to people, objects, and places, either in the physical world or in a fictional world of an author's creation. In a printed book, these references are usually identified by a separate index at the back of the book. However, with electronic texts, it should be possible for such references to be identified through suitable markup within the text itself.

6. *The historical nature of texts.* Printed and manuscript texts are historical objects that change over time. In most cases, there are multiple versions of a particular work that are likely to exhibit various differences and inconsistencies. Markup must

be able to show these variations and changes in a way that preserves their source and allows them to be compared and analyzed.

Types of Markup

The previous characteristics of texts are what a markup system needs to address if it is to provide an electronic representation of a physical text. In assessing existing markup systems against these goals, it is useful to identify the major types of markup that are embodied in such systems. In a seminal paper published in 1987, James Coombs, Allen Renear, and Steven DeRose identified four basic types of markup that are applicable in the electronic context: punctuational, presentational, procedural, and descriptive.

Punctuational Markup

Punctuation is so pervasive and familiar that it needs little ex-planation. The use of marks such as commas, colons, and periods (full stops) to provide syntactic information in sentences is univer-sally accepted and understood. However, certain areas of difficulty exist with regard to punctuation marks in an electronic context. Quotation marks, for example, can be rendered in various ways by different kinds of word-processing software. Some systems—such as Word 7.0—can distinguish between opening and closing marks for quotations. Some can distinguish single quotation marks from double. Hyphens, en dashes, and em dashes are a notoriously prob-lematic area, with most software unable to distinguish between them. The ambiguous use of some punctuation marks, such as peri-ods in abbreviations as opposed to at the ends of sentences, can also create problems.

In these difficult areas, it is possible—and often desirable—to replace punctuational markup with descriptive markup. Quotations might be indicated by the tags <quote> and </quote>, for instance. But, for the most part, punctuational markup is relatively straight-forward. Because it relates to the lowest level of the text—its phrases and sentences—it can be absorbed within one of the other types of markup and coexist with it, generally quite comfortably.

Presentational Markup

Presentational markup deals with a higher level of the text than punctuational markup does. Its purpose is to indicate the way in which the text should be presented to the reader. It identifies component parts of the text—paragraphs, pages, lists, notes, and the like—and marks them with such tools as line spacing, indentation, pagination, and sequential numbering. It also formats the text, using such devices as justification, font size, and bold or italic type styles.

Originally, this type of markup was added directly to the text by the author or copyist in the course of preparing a manuscript or typescript. Today, word-processing software does most of this work, inserting page numbers automatically, for example, or creating lists with bullet points. Indentation, spacing, and justification are set with rulers and tool bars. Styles and fonts are controlled with tool bars and drop-down menus. The presentational markup is inherent in the software itself; the author or copyist only needs to select from the choices it offers.

Procedural Markup

Procedural markup is closely related to presentational markup, in that its purpose is to show how the text should be formatted and presented. It expresses the markup as a series of codes that can be interpreted by the particular software being used as instructions for processing the text. In essence, this is a kind of computer programming language, aimed at formatting and presenting text as output.

In older text-processing software such as TeX and nroff/troff, these codes must be typed in by the editor or author. But most word-processing software can now automatically convert its files into formats that use procedural markup. This is especially the case with formats that are intended to be more interchangeable than proprietary word-processing software. Among the best-known are PostScript and RTF (Rich Text Format). PostScript is a programming language designed primarily for conveying printing instructions to laser printers. It uses such commands as:

```
/Helvetica findfont 12 scalefont setfont
45 458 moveto (Normal spacing) show
```

to produce 12-point Helvetica, with normal spacing (Adobe Systems, 1985:239). RTF is a file format developed by Microsoft for retaining a document's formatting instructions and exchanging them between different types of word-processing software.

Descriptive Markup

Descriptive markup differs from the other three markup types. Whereas they focus on the format and appearance of the text and convey instructions on its layout and printing, descriptive markup focuses on the content of the text. Particular elements within the text are identified using starting and ending tags, such as <quote> and </quote> before and after a quotation, respectively. This kind of markup may indicate the structure of the text as well as its content through the use of tags such as <body>, <head>, and <div>. Instead of labeling a section heading as "Times 18-point bold", it can be labeled as <div1><head> or something similar. No formatting is presupposed, but the function of the heading within the text is recorded and specified.

It is easy to confuse descriptive markup with procedural markup, and some authors have taken to using descriptive tags for procedural features. An example often seen on the World Wide Web is the use of <b> for bold, instead of <em> for emphasis. The former indicates how the text should be formatted and is therefore procedural markup. The latter indicates the purpose and function of the text and is therefore descriptive. This use of descriptive markup defeats its purpose since it is intended to separate the content of the text from the possible ways in which it can be formatted by different brands of software. If it is to be used successfully, this principle must be observed in a consistent manner. Otherwise, descriptive markup will lose its independence from specific kinds of formatting and presentational software.

Evaluating Different Types of Markup

Descriptive markup has several major advantages over the other main types. To begin with, it needs no maintenance. Once the text is marked up, it will not need revising to cope with new versions of software or the replacement of one kind of software with another. Nor will it need to be amended to allow for changes to formatting specifications issued by a publisher or a scholarly society. The source file can be separated from the changes that are continually taking place in the software environment.

Also crucial, descriptive markup can overcome the problems caused by incompatibility between different types of software. The text can be easily transported between systems, without the need for reformatting or rekeying. This is especially valuable in the publishing process, where authors may be using software which is incompatible with that used by their publisher. There may also be significant advantages for authors in being able to share documents with colleagues without the need to overcome incompatibilities in formatting. Other possible solutions to this need for portability are more clumsy, either requiring authors to produce camera-ready typesetting, or relying on less efficient techniques such as Optical Character Recognition (OCR) (see Chapter 2).

Descriptive markup is also "media neutral." In other words, it is not oriented toward an end product in a specific format. Other kinds of markup tend to be directed toward one medium, particularly print, and may need considerable reformatting for another. But files with descriptive markup can be reused for a variety of different formats—print, CD-ROM, networks—without any need to change the source files themselves. The reformatting is taken care of by specific software that processes these source files to produce the desired output. In this way, the formatting for output is kept separate from, and subsequent to, the markup.

With presentational and procedural markup, authors are obliged to focus on the appearance of the text as well as its content. This involves learning the formatting rules of a particular publisher or style guide, remembering and applying them at the appropriate time and place, and then relearning them when the rules change. In addition, with procedural markup, authors must remember the codes to be inserted at particular points in the text. All this effort can only distract authors from the process of composition, and may delay it considerably. None of this is necessary with descriptive markup. Authors need only indicate the function of particular parts of the text, not the way in which they should appear.

A further advantage of descriptive markup is that it embeds in the text a range of structural and semantic information that can be used for subsequent analysis of the text. In conjunction with suitable analytical software, searches of the text for particular words or phrases can be carried out within specific structural parameters. Occurrences of terms can be sought within quotations, for example, or within the critical apparatus. The hierarchical structure of the text can be displayed in a format similar to a table of contents, except that layers can be revealed or concealed at will. Alternative views of the text can be constructed, with different groups of elements displayed or hidden. Without such markup, the text is largely one-dimensional; it can only be searched as a whole or not at all.

In these ways, descriptive markup is far more satisfactory than the other types. They need more maintenance, and their portability is lower. They are generally aimed at a printed product and need to be reformatted for other media. They force the author to concentrate on the appearance of the text as much as its content, and they do not allow for sophisticated analysis of the text. It follows from this that descriptive markup is the best and most worthwhile solution for electronic texts which are intended to survive the continual changes in the world of software and which are designed to be studied and analyzed.

MARKUP-FREE TEXTS?

An alternative approach that is sometimes proposed involves so-called "markup-free" or "ASCII-only" texts. ASCII—the American Standard Code for Information Interchange—is a fundamental definition of the codes used by most computers to store internally such things as letters, numbers, and punctuation. The term "ASCII-only text" is a somewhat ambiguous and misleading one, but it generally refers to a file containing the raw text, with only the most basic markup: spacing, carriage returns, and punctuation. There is no formatting, no indentation or justification, and no indication of the structure of the text.

ASCII texts of this kind can be copied from one computer to another comparatively easily. Most brands of proprietary word-processing software can display and manipulate the text. But an

ASCII text on its own is of minimal scholarly value. It has no unambiguous way of indicating the structure of the text, nor of providing semantic information for the analysis and study of the text. Nor can it deal with the accents and diacritics of languages other than English, since these are not covered by the standard ASCII character set.

As Sperberg-McQueen points out, this is a misguided attempt to achieve texts with high portability and low maintenance. It rests on an "inadequate theory of texts, in effect claiming that the only essential part of a text is its sequence of graphemes" (1991:35). The result is a severely limited and impoverished text that has little immediate value for the scholar. The electronic texts produced by *Project Gutenberg* for distribution over the Internet have created considerable discussion and controversy for this very reason. Because they contain an ASCII text with no descriptive markup, they can only be read or searched in a simplistic way. They are the raw material for a more scholarly text, or one with more sophisticated presentation, but this will involve a good deal more work and the addition of some kind of descriptive or presentational markup. As well as *Project Gutenberg*, several of the other Web-based projects that aim to distribute electronic versions of well-known texts also use ASCII texts without markup. Among the better-known are the *Online Book Initiative* and the *Wiretap Online Library* (see Figure 1.1).

DESCRIPTIVE MARKUP SYSTEMS

HTML (HyperText Markup Language)

HTML, HyperText Markup Language, is the descriptive markup used for documents published on the Internet through the World Wide Web. HTML files themselves generally use only the basic ASCII character set and are, for the most part, independent of any proprietary software. They have several powerful features. They can embed within themselves "hypertext" links to other files, enabling the reader to jump to another HTML file stored anywhere in the world. They can refer to images and to other kinds of "multimedia"

FIGURE 1.1. *Wiretap Online Library:* William Law, *A Serious Call To a Devout and Holy Life* (Selection)

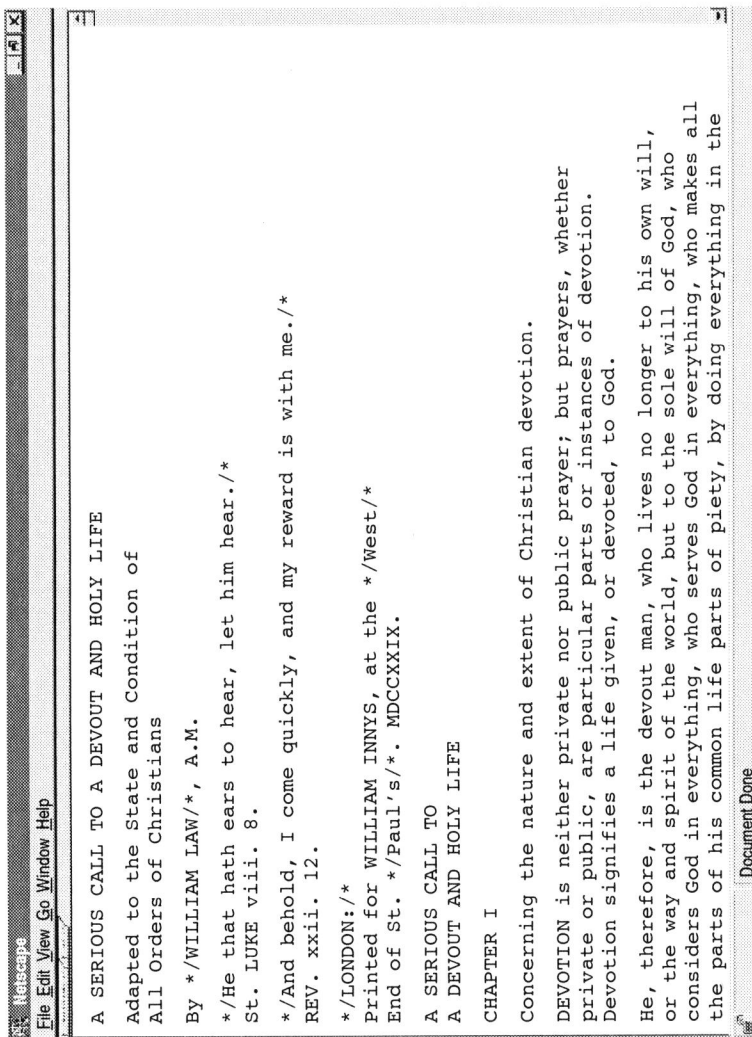

Netscape

File Edit View Go Window Help

A SERIOUS CALL TO A DEVOUT AND HOLY LIFE

Adapted to the State and Condition of
All Orders of Christians

By */WILLIAM LAW/*, A.M.

/He that hath ears to hear, let him hear./
St. LUKE viii. 8.

/And behold, I come quickly, and my reward is with me./
REV. xxii. 12.

/LONDON:/
Printed for WILLIAM INNYS, at the */West/*
End of St. */Paul's/*. MDCCXXIX.

A SERIOUS CALL TO
A DEVOUT AND HOLY LIFE

CHAPTER I

Concerning the nature and extent of Christian devotion.

DEVOTION is neither private nor public prayer; but prayers, whether
private or public, are particular parts or instances of devotion.
Devotion signifies a life given, or devoted, to God.

He, therefore, is the devout man, who lives no longer to his own will,
or the way and spirit of the world, but to the sole will of God, who
considers God in everything, who serves God in everything, who makes all
the parts of his common life parts of piety, by doing everything in the

Document Done

files. If their hypertext structure is well-designed, they can allow browsing by tables of contents, as well as the linear reading offered by basic ASCII files. HTML files can also be searched for specific strings of letters and other characters by various types of searching software or "search engines."

The original version of HTML was developed by Tim Berners-Lee at the CERN Laboratories in Switzerland in the early 1990s. When combined with the HyperText Transport Protocol (HTTP) and a system for standardized naming of computer resources (URLs, or Uniform Resource Locators), it provided the basic architecture needed to transform the Internet into the World Wide Web and was largely responsible for the amazing growth in the Web's popularity. The appearance of graphical software—initially Mosaic, but later Netscape and Internet Explorer—for browsing the Web and viewing HTML documents was a major element in this process.

An important reason for the success of HTML is its simplicity. The earliest versions of HTML contained fewer than forty different elements. In practice, only a handful of these were needed to mark up a document. It was possible to publish a document on the Web that contained only a small number of tags: <HEAD>, <BODY>, a couple levels of headings using <H1> and <H2>, paragraphs <P>, hypertext links <A>, and perhaps an unordered list <UL> with list items <LI>. Images could easily be embedded with the <IMG> tag. The result was a hypertext document that could be linked into the Web, with comparatively little effort or knowledge.

Since then, HTML has become increasingly complex and elaborate, partly through the work of official committees and standards groups, and partly through the work of software developers such as Netscape and Microsoft. One of the major developments has been the introduction of the <FORM> tag, which allows the creation of input forms. The person reading a page can type in data, particularly search queries, and send them to a distant computer for processing. A program or script is then invoked on the "home" server, which will carry out a search and return the results as a Web page to the user. This approach is employed by all the search services on the Web, such as Alta Vista, Hot Bot, and Lycos. The pages produced in response to a search query such as this do not usually exist as distinct files. The underlying structure is generally a relational database that is searched by the query contained in the form in the original HTML page. An HTML page, or pages, is constructed automatically by the software in response to the specific request. This kind of facility turns HTML pages from passive documents into active and dynamic ones.

Another occasion on which HTML pages are constructed in response to a request is when the source files are stored in a different kind of markup, usually more elaborate than HTML. Once again, programs or scripts are invoked by the user's request, which will automatically translate documents from another format into HTML. The Electronic Text Center in the University of Virginia Library uses this kind of approach to serve up HTML versions of texts that are actually stored in much more detailed markup (Seaman, 1994).

Another important recent feature of HTML is its ability to link small computer programs into an HTML page. These so-called applets, written in a programming language called Java, produce an illusion of action and movement—words moving across the screen, a globe or a head turning around in an apparently three-dimensional field. They tend to decorate the screen rather than add to the value of its content. The important characteristic of Java programs is that they are sent from the Web server to the remote user's computer, and run there with the browser software. They are linked into the HTML page with a tag called <APPLET>. A different approach, but with similar results, is adopted by JavaScript. This method actually includes the ASCII text of a program in the HTML page, using the <SCRIPT> tag.

The development of HTML over the past four years has followed a somewhat convoluted path. A considerable amount of confusion and incompatibility was caused by the activities of the leading software companies, which competed to introduce new tags and features into their browsers without waiting for the official version of HTML to support them. Tables, for instance, were used by Netscape long before they were incorporated into version 3.2 of HTML, which was only adopted as a recommendation of the World Wide Web Consortium in January 1997. These so-called "tag wars" of the mid-1990s meant that Web pages marked up in accordance with one company's specifications might not be viewable with another company's browser. Of the various nonstandard tags in existence, the most notorious are "frames," which split the screen into two or more separate "pages" or windows (Wusteman, 1997).

An end to these proprietary extensions may have been reached with the latest version of HTML, 4.0, which became a recommendation of the Consortium in December 1997. This is specified in three versions:

- HTML 4.0 Strict
- HTML 4.0 Transitional
- HTML 4.0 Frameset

The Frameset approach finally provides official endorsement of the use of frames to partition the browser window. The Transitional approach and the Strict approach differ mainly in the way they handle presentational features. HTML 4.0 Transitional continues the practice of including presentational markup in the HTML tags themselves. In contrast, HTML 4.0 Strict separates the presentational markup and handles it through the Cascading Style Sheet language, or CSS. Only the latest versions of browser software will be able to handle CSS, which relies on links from the HTML page to a style sheet describing how particular elements should appear on the screen or in print. This approach gives the Web designer much greater control over the appearance of a Web page—its fonts, margins, line spacing, borders, colors, and so on. It promises to rectify one of the biggest areas of inconsistency and confusion with HTML: its indiscriminate mixing of descriptive and presentational markup.

Cascading Style Sheets are a key component in what has been labeled "dynamic HTML." This term refers to HTML pages with dynamic content rather than the static content of basic HTML markup. Another important part of dynamic HTML is JavaScript, but perhaps the most significant new feature in this area is the Document Object Model. This aims to specify a generic approach for manipulating HTML with scripts and programs such as Perl and Java. The Document Object Model defines a set of methods and data types that are independent of any particular programming language or software and hardware platform.

Despite all the ups and downs of the "tag wars," HTML has been remarkably successful. It has been so widely adopted as a markup language that there are now hundreds of millions of HTML pages on computers around the world. In April 1998, there were over 2.2 million Web sites, with hundreds—if not thousands—of HTML pages on each. But the limitations of HTML as a scholarly tool remain considerable. For all the work on forms, applets, and frames, it still makes possible only a comparatively low level of descriptive markup for the structure and content of a text. From this perspective, most HTML files contain little more than a basic structure of paragraph divisions with headings. HTML has been used largely as a presentational tool rather than a structural, analytical one (Flynn, 1997b).

HTML is used as the preferred markup language for various electronic texts available on the World Wide Web. These tend to be personal collections rather than institutional projects or digital libraries. Many of the classical and medieval texts assembled by James J. O'Donnell at the University of Pennsylvania are in HTML markup, for instance. The *Online Literature Library* also relies on HTML, which is applied to ASCII texts from *Project Gutenberg* and the *Wiretap Online Library* (see Figure 1.2).

SGML (Standard Generalized Markup Language)

HTML is a specific and very restricted application of the international standard for markup languages: SGML, the Standard Generalized Markup Language (International Standard ISO 8879). SGML is a metalanguage that sets out the principles of a notation system for "representing documents that are acted on by text processing applications" (Goldfarb, 1990:123) and prescribes rules for constructing specific markup languages for particular applications.

From work on generic coding in the later 1960s, sponsored by the Graphic Communications Association, a research team at IBM led by Charles Goldfarb developed a Generalized Markup Language (GML). This later became the basis for an international standard, which was adopted, as SGML, in October 1986. An amendment to the standard was approved in July 1988. Standards for specific markup languages in the publishing industry and the defense industry, derived from SGML, were adopted in 1987 and 1988, respectively. In the subsequent ten years, SGML has become widely used for processing texts and documents by major compa-

FIGURE 1.2. *Online Literature Library:* Lewis Carroll, *Alice's Adventures in Wonderland* (Selection)

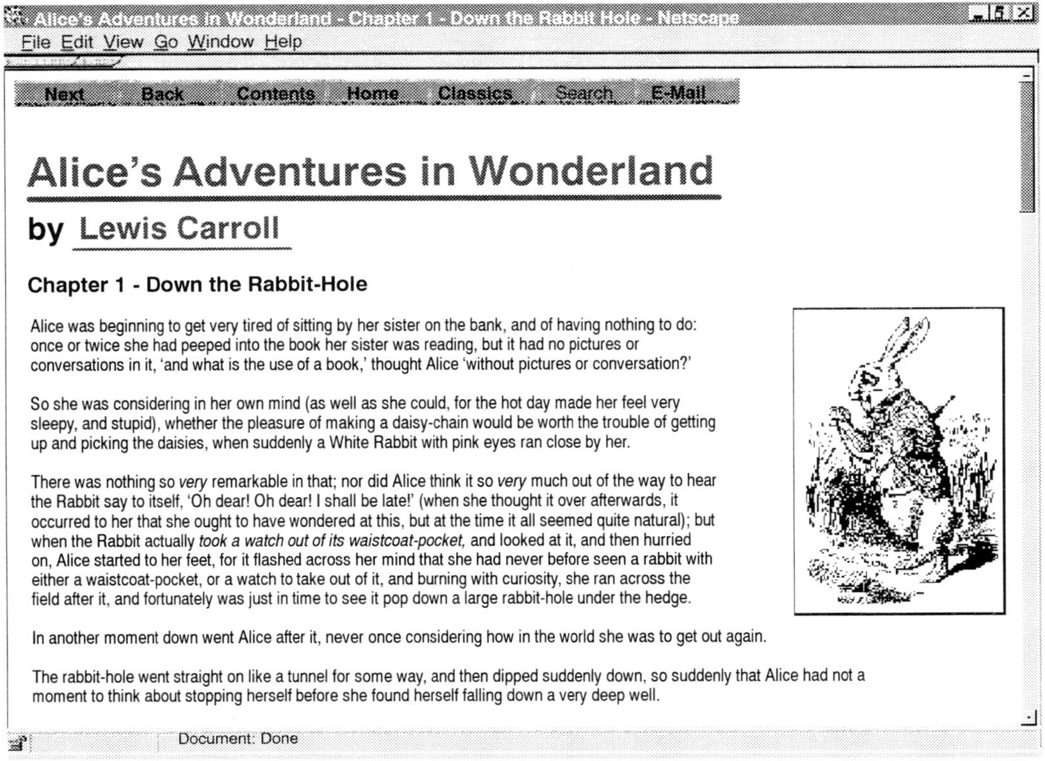

Alice's Adventures in Wonderland - Chapter 1 - Down the Rabbit Hole - Netscape

File Edit View Go Window Help

Next Back Contents Home Classics Search E-Mail

Alice's Adventures in Wonderland

by Lewis Carroll

Chapter 1 - Down the Rabbit-Hole

Alice was beginning to get very tired of sitting by her sister on the bank, and of having nothing to do: once or twice she had peeped into the book her sister was reading, but it had no pictures or conversations in it, 'and what is the use of a book,' thought Alice 'without pictures or conversation?'

So she was considering in her own mind (as well as she could, for the hot day made her feel very sleepy, and stupid), whether the pleasure of making a daisy-chain would be worth the trouble of getting up and picking the daisies, when suddenly a White Rabbit with pink eyes ran close by her.

There was nothing so *very* remarkable in that; nor did Alice think it so *very* much out of the way to hear the Rabbit say to itself, 'Oh dear! Oh dear! I shall be late!' (when she thought it over afterwards, it occurred to her that she ought to have wondered at this, but at the time it all seemed quite natural); but when the Rabbit actually *took a watch out of its waistcoat-pocket,* and looked at it, and then hurried on, Alice started to her feet, for it flashed across her mind that she had never before seen a rabbit with either a waistcoat-pocket, or a watch to take out of it, and burning with curiosity, she ran across the field after it, and fortunately was just in time to see it pop down a large rabbit-hole under the hedge.

In another moment down went Alice after it, never once considering how in the world she was to get out again.

The rabbit-hole went straight on like a tunnel for some way, and then dipped suddenly down, so suddenly that Alice had not a moment to think about stopping herself before she found herself falling down a very deep well.

Document: Done

nies in the defense, aerospace, legal, publishing, and financial industries.

SGML and the markup languages derived from it have several important advantages (Hockey, 1996:6-7):

- Texts encoded in this way can easily be moved from one platform to another.
- They are entirely independent of any particular brand of computer program or equipment.
- They exist as plain ASCII files.
- The encoding can be much richer and more flexible than other markup schemes.
- The encoding can be incremental, with different users adding their own layers of markup.
- Analytical material can be included as a specific component of the markup.
- Sophisticated mechanisms for cross-referencing can be embedded in the text.

This combination of portability, independence, and sophistication makes SGML the best model for marking up scholarly texts. This is reinforced by the international application of the SGML standard and by the availability of a wide variety of software designed to handle SGML-based texts.

SGML treats a text as a collection of interrelated objects. The various component objects in the structure of the text are described in terms of their function within the text. In SGML, these objects are known as *elements* and are indicated within the text by a start tag and an end tag:

<line>On either side the river lie</line>

Here, <line> is an element used to mark (obviously enough) a line within a poem. The end tag is differentiated from the start tag by the addition of the forward slash.

Each element may have one or more *attributes*, which provide more specific information about different occurrences of the element and analyze it in a more detailed way:

```
<line n=1>On either side the river lie</line>
<line n=2>Long fields of barley and of rye,</line>
<line n=3>That clothe the wold and meet the sky;</line>
```

Here, each occurrence of the <line> element in the poem is distinguished by the addition of an attribute "n," which specifies the number of each line within the poem as a whole. This attribute provides useful additional information about the element that can be used in searching or displaying the poem.

Elements and their attributes are intended to identify the structural components of a text. Another important component of SGML-based markup languages is the *entity*, which is defined as "a collection of characters that can be referenced as a unit" (Goldfarb, 1990:129). Entities provide a shorthand method of referring to something more complex than a character. Their most common uses are to:

- identify characters that cannot be keyed directly;
- provide a shortcut for a frequently used and lengthy piece of text;
- embed the results of a processing instruction into a document; and
- incorporate external resources, such as data files, into a document.

Most entities are indicated in the text by an ampersand before and a semicolon afterward. The French word for theater, with an acute accent on the "e" and a circumflex on the "a," appears like this:

le théâtre

An external entity might look like this:

```
<chapter>
&chapter1;
</chapter>
```

where the full text of Chapter 1 is stored externally in a separate file.

For marking up any specific text, the SGML standard prescribes that elements, attributes, and entities must be used in a formally defined and consistent manner. There are three essential components:

- The SGML declaration
- The Document Type Definition
- The document itself ("the document instance," in SGML terminology)

The SGML declaration specifies the character set and the encoding syntax used in the document. It is usually already present in default form in software designed for processing SGML-based texts and does not need to be specified in relation to individual documents.

The Document Type Definition (DTD) *does* need to be specified, however. A DTD can be thought of as a markup language constructed in accordance with the SGML standard, or as the set of rules that governs the markup of a particular text. SGML regards each document as belonging to a class or group of similar documents, known as a "document type." In formal terms, a specific document belongs to its document type when it observes the formal markup rules for that class of documents. The DTD defines which elements can be used in marking up a document, which attributes are valid for each element, and which entities are allowable. In particular, it sets out the hierarchical structure for marking up documents, by defining the context within which each element can be used.

This contextual information is called a *content model:*

<!ELEMENT poem - 0 (title?, (stanza+ | couplet+ | line+)) >

In this example, the content model for the element "poem" is contained within the round brackets. It can be interpreted as meaning that the element "poem" may contain the subsidiary element "title" (the optional nature is indicated by the question mark), but *must* contain one of the elements, "stanza," "couplet," or "line" (the alternatives are shown by the vertical bars). The plus sign means that there may be multiple occurrences of these elements—otherwise only a single occurrence would be permitted. The order in which the subsidiary elements are listed is important; this example indicates that if a "title" element is present, it must occur before the other subsidiary element(s). The other information conveyed by this definition, in the two characters before the content model, relates to what is called minimization. In this case, the hyphen means that the start tag <poem> must be used to mark the beginning of this element in the text, but the end tag </poem> does not need to be used to mark the end of the element, which can be assumed from the context.

The content model is used as the basis for marking up the text—in this case, of the poem:

> <poem><title>The Lady of Shalott</title>
> <stanza n=1>
> <line n=1>On either side the river lie</line>
> <line n=2>Long fields of barley and of rye,</line>
> <line n=3>That clothe the wold and meet the sky;</line>
> . . .
> <line n=9>The island of Shalott.
> </stanza>
> <stanza n=2>
> <line n=10>Willows whiten, aspens quiver,</line>

and so on, through to the end of the poem. Here, the content model for the <poem> element begins with a title, followed by a series of numbered stanzas. Each stanza, in its turn, contains a series of numbered lines. The line numbering runs consecutively through the poem. Although this is a simple example, lengthy and complicated texts require a very detailed and elaborate content model to express their hierarchical and linear structures.

A DTD must contain definitions such as this for all the elements that may occur in valid documents of that type. It must also contain definitions of all the valid attributes that can be used with each element, expressed in the form:

> <!ATTLIST poem   id   ID   #IMPLIED
>   status (draft/published) draft   >

This attribute list means that the element "poem" may have two attributes, "id" and "status." The former does not have to occur (indicated by the notation #IMPLIED), but takes the form "ID" when it does occur. The latter must be either status = "draft" or status = "published," but defaults to "draft."

Entities must also be defined in the DTD, in the form:

```
<!ENTITY      lt          "<" >
<!ENTITY      chapter 1   SYSTEM "\user\docs\chapter1.doc">
```

The first line means that whenever the entity reference "<" appears in the text, it will be interpreted as a left-angled bracket sign. An entity of this kind is necessary because the angled bracket characters are reserved for special uses in SGML. The second line means that when the entity reference "&chapter1;" appears in the text, the contents of a file called "chapter1.doc" will be incorporated at that point. The entity declaration records that this file is present on the same computer and gives its directory path.

For marking up texts in the humanities, there is a highly detailed and sophisticated DTD known as the TEI. It was developed between 1987 and 1994 by representatives from eighteen different learned societies, who formed what was called the Text Encoding Initiative. The official guidelines that make up the TEI were published in April 1994 (Sperberg-McQueen and Burnard, 1994). They are intended to provide "a standard format for data interchange in humanities research" and to set out "principles for the encoding of texts in the same format" (Ide and Sperberg-McQueen, 1995:6). The specific design goals for the TEI were formulated in the following terms (Sperberg-McQueen and Burnard, 1994:17). The guidelines aim to be:

- adequate for representing the textual features needed for research;
- simple, clear, and concrete;
- easy to use without specialized software;
- sufficient for the rigorous definition and efficient processing of texts;
- capable of being extended by their users; and
- conformant to standards.

In practice, the TEI defines about 400 different elements. The crucial factor, however, is the flexibility with which the elements can be used and combined. Within the TEI framework, there are several modules that are described as "DTD fragments," or tag sets (Sperberg-McQueen and Burnard, 1994:37-41). They can be combined in various prescribed ways within the main TEI DTD. These modules fall into three groups:

- "Core" tag sets: standard components that are used in all forms of the TEI DTD
- "Base" tag sets: eight different sets of tags designed to handle specific forms of material
- "Additional" tag sets: extra tags for specific purposes, which can be used with each of the base sets

Using the TEI involves combining the core tag sets with one of the base tag sets and any of the additional tag sets, an approach jokingly dubbed the "Chicago pizza model" (Sperberg-McQueen and Burnard, 1994:27). While the core tag sets contain common or generic elements and the bibliographic elements making up a file header, the base tag sets provide elements relevant to a specific type of document: prose, verse, drama, and so on. The additional tag sets include such things as hypertext linking, segmentation, apparatus, tables, and graphics.

For those who do not need or want the full power and complexity of the TEI DTD, a simplified and considerably abridged version has been devised, known as TEI Lite (Burnard and Sperberg-McQueen, 1995). Even more minimalist is what is known as "Bare Bones TEI," which contains a "very very small subset" of the TEI (Sperberg-McQueen, 1994). TEI Lite is an important and valuable tool, used by the Oxford Text Archive for many of its texts, for instance, but Bare Bones TEI is intended only for illustrative purposes, not as a working DTD in its own right.

For encoding the structure of a text, the TEI in all its versions provides a comparatively simple and straightforward default. The text is divided into three main sections: a body, a front, and a back. The latter two are optional. These divisions are indicated by the elements <BODY>, <FRONT>, and <BACK>, which occur within the <TEXT> element. An alternative approach, aimed particularly at anthologies consisting of various separate texts, is to enclose each text within a <GROUP> element. In the body of a text, the major structural divisions are marked by a <DIV> element. Different types of structures can be shown by the type attribute, for example, <DIV TYPE="chapter">. Where the hierarchy of divisions and subdivisions needs to be indicated, elements of the form <DIV0>, <DIV1>, <DIV2>, and so on can be used for each successive layer of this structure.

An important general feature of the TEI encoding scheme is the use of global attributes that apply to every element in the scheme. One of these attributes, id, allows each element occurring in a text to be given a unique identifier. This makes possible extensive cross-referencing and linkages within the text, interweaving a hypertextual web across the formal hierarchical structure of the text. Other global attributes include a numbering system for each element, which is not necessarily unique, using the attribute n, and the ability to specify the language of the content of an element using the attribute lang.

The TEI also contains many useful features for encoding particular aspects of literary texts (Robinson, 1994:77-90). Graphic images within a manuscript or printed text can be identified with a <FIGURE> tag, which is able to refer to separate image files and to tell the processing software what type of file is involved. Abbreviations in the original text can be transcribed using the <ABBR> tag, with the expansion given in an attribute called expan. Alternatively, the expansion can be preferred as the primary transcription, using the <EXPAN> tag, with an abbr attribute to record the abbreviation. Alternative transcriptions and readings are given within the <RDG> tag and marked as part of the apparatus <APP>. Corrections are indicated by the <SIC> tag and a corr attribute or by the reverse approach—the <CORR> tag and a sic attribute. Regularized or normalized spelling is distinguished from the original spelling by the <REG> tag and an orig attribute or by the reverse approach—the <ORIG> tag and a reg attribute.

Also of importance is the TEI file header, which is designed to "document the bibliographic description of the electronic text and its source" (Giordano, 1995:76). It prescribes three major pieces of information about the text: its title, its publisher or distributor, and the source from which it was derived. Optionally, it may also document the practices used in the encoding of the text, as well as providing classificatory information about the text itself, and recording changes made to the electronic file over time.

The TEI DTD is widely used in electronic text projects, particularly those of a more ambitious and long-term nature. Among these are the *Canterbury Tales Project*, the *Electronic Piers Plowman*, the collections at the University of Virginia, and those which form the Humanities Text Initiative at the University of Michigan. Other collections, notably those of the British publishing company Chadwyck-Healey Ltd., use an SGML DTD that is similar, but not identical, to the TEI. In this context, the decision of the *Hartlib Papers* project not to use SGML-based markup, and to publish in Microsoft Word 4.0 files instead, is interesting. It has been justified on the following grounds (Leslie, 1993:47-48):

- SGML and the TEI were insufficiently developed when the project began in the late 1980s.
- The time required for marking up an archive of twenty million words would have delayed the completion of the project by several years.
- SGML-aware text-retrieval software was not as powerful as the TOPIC software used by the project.

Although perhaps a result of their bad timing, the *Hartlib Papers* suffer in comparison with later projects based on SGML and the TEI. The files are in a proprietary format that is now obsolete. Bringing it up to date, perhaps for delivery over the World Wide Web, would require extensive reformatting.

The TEI guidelines are not the final word on electronic text encoding and interchange for research in the humanities. However, they are a remarkable achievement and have undoubtedly provided a definitive framework for electronic texts, which can be further developed and refined in the future. (See Figure 1.3 for an example of TEI markup.)

XML (Extensible Markup Language)

A more recent development, which is potentially of great significance for text markup, is the Extensible Markup Language (XML). A project of the World Wide Web Consortium, XML first emerged

FIGURE 1.3. Dante, *La Divina Commedia* (Selection)

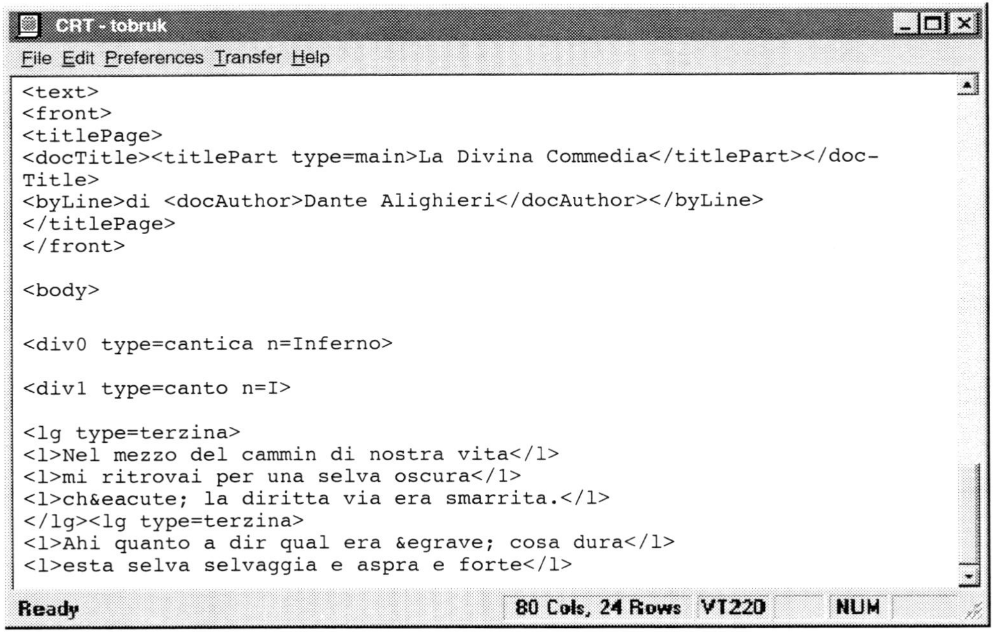

```
CRT - tobruk
File  Edit  Preferences  Transfer  Help

<text>
<front>
<titlePage>
<docTitle><titlePart type=main>La Divina Commedia</titlePart></doc-
Title>
<byLine>di <docAuthor>Dante Alighieri</docAuthor></byLine>
</titlePage>
</front>

<body>

<div0 type=cantica n=Inferno>

<div1 type=canto n=I>

<lg type=terzina>
<l>Nel mezzo del cammin di nostra vita</l>
<l>mi ritrovai per una selva oscura</l>
<l>ch&eacute; la diritta via era smarrita.</l>
</lg><lg type=terzina>
<l>Ahi quanto a dir qual era &egrave; cosa dura</l>
<l>esta selva selvaggia e aspra e forte</l>

Ready                                    80 Cols, 24 Rows  VT220   NUM
```

TEI Lite markup by James Tauber for the University of Western Australia Library.

toward the end of 1996 and was further elaborated and specified during 1997. The XML 1.0 specification was officially released as a recommendation of the Consortium in February 1998. The main purpose of XML is to find a middle ground between HTML and the full SGML standard for presenting resources over the Web (Dougherty, 1997; Flynn, 1997a). From the SGML point of view, XML is an abbreviated version that omits various complicated and less-used aspects of the standard. But it remains, similar to SGML, a metalanguage rather than a markup language in itself. Specific markup languages must be developed in order to apply XML to specific documents.

Unlike SGML, though, these specific markup languages can be designed in two different ways:

- Use an SGML-like Document Type Definition (DTD), which can then be applied to what is called a valid document.
- Create what is called a well-formed document, which does not need a DTD but must follow certain simple rules about the occurrence of elements and attributes.

Both types of documents must be consistent and parsable, but the parsing process works differently in each case. Well-formed documents are not validated by the parser software against a DTD, merely checked for consistency of structure. For the sharing and transmission of similar types of documents, however, a valid document that conforms to a standard DTD will be much more useful than a well-formed one.

XML is designed for use with the next generation of Web-browser software. It is intended to address and overcome various deficiencies in HTML, especially its fixed tag set, its inconsistencies and incompatibilities, and its simplistic approach to hyperlinks. A key feature of XML is its ability to permit Web authors and designers to create their own set of elements for specific markup, instead of having to rely on the fixed set of elements offered by HTML. XML will allow individuals or groups to use customized markup for their Web pages. Another area in which XML is expected to be particularly useful is in devising formats for metadata (bibliographic information about Web resources). It is also designed to be much more hospitable than HTML to programming languages such as Java.

Two closely related developments are XSL and XLL. XSL is the proposed Extensible Stylesheet Language, currently being assessed by the World Wide Web Consortium. If adopted, it will define ways of creating style sheets for XML documents to control their appearance and formatting when viewed with browser software. XLL is the Extensible Linking Language. Also under discussion at present, it will provide a much richer range of linking mechanisms than HTML does. These include the ability to link to a small part of a large document and to display only that part, as well as links that work in both directions (instead of the one-way approach of HTML links). A single link to several different resources, presented as a menu, will also be possible.

In all these ways, XML offers a much more sophisticated approach to publishing material on the Web, with only a fraction of the complexity of SGML. The reaction has been enthusiastic. Major software companies such as Microsoft, Netscape, and Sun have agreed to support XML, as have most of the firms active in the SGML software market, such as Inso and SoftQuad. Software for creating, validating, and publishing documents marked up in accordance with XML is already under development.

At present, there are no examples of scholarly texts that use XML. However, it should not be too long before such texts begin to appear. An XML version of TEI Lite is already available, and work is underway to translate the full TEI into XML. It seems likely that XML will complement SGML-based markup similar to the TEI, rather than superseding it. The TEI will continue to be used and developed as the de facto standard for the scholarly encoding of texts.

FURTHER READING

On the theory of markup, see the articles by Coombs and others (1987) and Renear (1992).

There is a rapidly growing range of books devoted to SGML. Robinson (1994) provides a succinct introduction. Perhaps the best detailed account for the novice is von Hagen (1997). A useful reference tool for SGML is Bradley (1997).

On the relationship between SGML and HTML, see Bryan (1997) and Rubinsky and Maloney (1997).

The definitive guide is *The SGML Handbook* (Goldfarb, 1990), which contains the full text of the SGML standard (ISO 8879) with extensive annotations and a fairly technical overview of the concepts underlying SGML.

Books that examine SGML from the perspective of implementing it as part of a document management system include Alschuler (1995) and Travis and Waldt (1995). Ensign (1997) deals with commercial applications of SGML, but is still of interest.

There are many sources of information about SGML on the World Wide Web. Among the most important are the Usenet news group comp.text.sgml and Robin Cover's definitive SGML Web page. Other useful sites include SGML University, the SGML Open Home Page, Charles F. Goldfarb's Web site, and SGML on the Web. The International Standards Organization provides official material relating to the SGML standard.

The *Guidelines* for the TEI (Sperberg-McQueen and Burnard, 1994) contain an introduction to the principles of this markup scheme, as well as an exhaustive list of its component parts. They also contain a "gentle introduction to SGML." The volume of essays on the TEI, edited by Ide and Véronis (1995), includes discussion of general issues, as well as analyses of the application of the TEI to specific types of text. Robinson (1994) gives a good short account of the practicalities of the TEI.

There are numerous books on HTML markup. Graham (1996) is one of the better ones. For Cascading Style Sheets, see Lie and Bos (1997).

On XML, see Mace (1998), Light (1997), St. Laurent (1998), and Connolly (1997). There is also a daunting array of Web-based material. Among the more useful sites are Arbortext's SGML and XML Resources, James Tauber's XML site, and the What is XML? site.

The Web site of the World Wide Web Consortium (W3C) provides official material on HTML and XML, as well as SGML.

Web growth statistics can be found at the *Hobbes Internet Timeline* site.

WEB SITES

Arbortext's SGML and XML Resources:
http://www.arbortext.com/linksgml.html

Charles F. Goldfarb's Web site: http://www.sgmlsource.com/

Comp.text.sgml (Usenet news group)

Hobbes Internet Timeline:
http://info.isoc.org/guest/zakon/Internet/History/HIT.html

International Standards Organization: http://www.ornl.gov/sgml/

James J. O'Donnell: http://ccat.sas.upenn.edu/jod/

James Tauber's XML site: http://www.jtauber.com/xml/

Online Book Initiative: http://world.std.com/references/obi.html

Online Literature Library: http://www.literature.org/

Project Gutenberg: http://promo.net/pg/

Robin Cover's SGML Web page:
http://www.sil.org/sgml/sgml.html

SGML on the Web:
http://www.ncsa.uiuc.edu/WebSGML/WebSGML.html

SGML Open Home Page: http://www.sgmlopen.org/

SGML University: http://www.sgml.com/

TEI: http://www.uic.edu/orgs/tei/

TEI Lite in XML: http://www.loria.fr/~bonhomme/xml.html

What is XML?: http://www.gca.org/conf/xml/xml_what.htm

Wiretap Online Library: http://www.area.com/

W3C (World Wide Web Consortium): http://www.w3.org/

Chapter 2

Creating an Electronic Text

Once a markup system has been chosen as the basis for encoding an electronic text, the essential next step is to create or capture an electronic version of that text. There are various different ways of carrying out this process. The most obvious is to key in the text from scratch; this will require decisions about quality control and rules for ensuring a consistent transcription. Where an electronic version of the text already exists, it may be possible to reuse it. This will depend on the quality and provenance of the existing version and on the relationship between its format and the format intended for the new version.

Scanning in the text as an image is another method of creating an electronic version. There are a range of important issues connected with image digitization, particularly in relation to the resolution and compression of images. Choosing between different file formats is also crucial. If the scanned image is to be turned into a text file, this will rely heavily on the degree of effectiveness of Optical Character Recognition software.

In this chapter, we examine the main ways of carrying out the process of creating an electronic text: keying it in, reusing an existing text, or scanning the text. We also discuss and compare text files and image files as formats for distributing an electronic text.

KEYING THE TEXT

If the text does not exist in electronic form already, the most obvious way of creating an electronic version is simply to type it into an electronic file. This may sound straightforward, but there are several important factors to be considered that will have a critical effect on the resulting text.

The Software Used to Key in the Text Must Produce a File That Is Compatible with the Desired Markup System of the Final Version of the Text

If the only markup required is so-called "plain ASCII" or a proprietary word-processor's format, the choice of software for input is relatively straightforward. Use your favorite word-processing software and save the file in the appropriate format.

For more sophisticated and portable markup, especially of the SGML kind, the decision about software is rather more complicated. A choice must be made between two different approaches to keying the electronic text:

- Create a raw text first, and add the markup later.
- Create the text directly with SGML-based markup.

It is not necessarily the case that the first electronic state of the text must contain the final markup. If the TEI is being used as the ultimate standard, for instance, the typist does not need to insert the appropriate TEI tags while keying in the text. Nor does the print or manuscript copy from which the typist is working need to be annotated with all the TEI markup before the typing begins. It is possible to add this kind of markup to an existing raw electronic version of the text, as a later and separate step.

The critical point is that the initial electronic file must be capable of being converted systematically, and without too much difficulty, into the chosen markup system. If the text is initially being created as a plain text file using common word-processing software, it is important that structural features of the text are easily identifiable and separable from the body of the text, to simplify later markup. This subsequent addition can often be done in a partly automated way, by using find-and-replace functions of word-processing software such as Microsoft Word. For a more sophisticated approach to this kind of task, UNIX text manipulation utilities such as Sed and Awk can be used, as can simple programming or scripting languages such as Perl. A useful account of a novice's approach to this process is given by James Tauber, who describes how he employed a range of these tools to add TEI-based markup to a raw electronic text (Tauber, 1995).

An alternative approach is to key in the initial electronic version of the text with the final markup already present. For HTML-based markup this can be done by using word-processing software that "understands" HTML and will embed HTML coding at the appropriate points in the text. Microsoft Word, for instance, comes with an application called Internet Assistant, which is designed to perform this translation to HTML. A potential problem with this approach is that the HTML markup itself is hidden from view. Another possibility is to use HTML authoring and editing software such as Adobe PageMill, Microsoft FrontPage, or SoftQuad's HoTMetaL Pro. These are sophisticated commercial products with an extensive range of features. There are also numerous free or shareware packages for this purpose, though these are more limited in their functions. HTML authoring software is examined in more detail in Chapter 3.

For more complex SGML-based markup, such as the TEI, there are so-called SGML editors that are designed for the composition of text directly into the fully marked-up version. Among the better-known of these are Arbortext's Adept Editor, SoftQuad's Author/ Editor, and Grif's SGML Editor. These have been used successfully for a range of scholarly projects in electronic texts. There are also SGML add-on packages for WordPerfect and Microsoft Word, such as Near & Far Author and Microsoft's own SGML Author for Word. These tend to be aimed at commercial applications rather than scholarly ones, and some investigation of their effectiveness with the TEI DTD would be highly desirable. A more scholarly application is Peter Robinson's Collate, which is designed for the input and collation of up to 100 versions of a text. The collation may then be output in TEI or HTML markup. Software tools of this kind are discussed in more detail in Chapter 3.

Some HTML and SGML authoring software works by offering menu choices of tags, while inputting a document, and allowing a choice to be made. Other programs automatically apply the "appropriate" markup, which is effectively hidden from the user. This latter type is designed for nonspecialist users and is unlikely to be helpful or sophisticated enough for complex scholarly schemes such as the TEI. This approach can be particularly frustrating to users who have some familiarity with the markup scheme involved because they are prevented by the program from editing the tags and attributes directly.

It may be preferable to ignore this kind of software and to key in the SGML markup as part of a plain text file, using basic word-processing or text-editing software. This will require either that the printed or manuscript text is annotated with the appropriate markup before keying begins or that the markup is added by the person doing the keying, as an integral part of the process of transcription. The Emacs text editor, in particular, has an SGML mode that will also parse the text against the appropriate DTD. Only with Unix text-editing software can you be sure that no "invisible" formatting or other special characters are being inserted into the text by the program.

Choosing between these two basic approaches—keying the text without SGML-based markup or composing it directly in an SGML-conformant form—will depend on the particular circumstances of the project. The characteristics of the text itself should be the first consideration in such a choice. If the text is irregular and complex, requires specialized knowledge to read, or contains features that cannot be easily indicated in a raw electronic version, then some kind of direct SGML composition or transcription may well be necessary. The direct employment of specialized staff is probably required to carry out such a task.

On the other hand, if the text is regular in structure and straight-forward in appearance and content, it may be cost effective to con-tract out the work. Some data entry bureaus now offer a total package that includes SGML or HTML in the cost of keying a text. As well as carefully comparing the cost of this service, it is essential to estimate the probable success of such a process. It is only likely to work well with consistent and easily legible texts in the English language. Most bureau services of this kind are aimed at the business market, with its comparatively regular and structured documents, and may rely on Optical Character Recognition (OCR) processes to reduce costs. Complex and irregular humanities texts are unlikely to be well-handled by such services. At the very least, the cost will be signifi-cantly above the standard rates quoted in price sheets. In many cases, it will be preferable to employ specialized staff directly rather than attempting to contract out the work to a bureau.

Keying the Text in Once May Not Be Enough

Even when a text exists in only a single printed or manuscript version, it may not be sufficient to key the text in once. A common practice to ensure the accuracy of the initial typing is to have the same text keyed in two or more times, into separate files. These files are then matched automatically to identify differences and discrepancies that can be resolved by checking against the original source. Commercial bureaus offer this kind of service, and the cost of keying is directly related to the number of times a text is keyed and matched. Single-pass keying, at a minimum accuracy rate of 95 percent, is likely to cost 50 percent less than double-pass keying, which should reach 99.95 percent accuracy. A third pass, to raise the level to 99.995 percent, will probably increase the cost by another 50 percent.

Keying the text in once will reduce the cost of this initial stage of the process. However, the amount of time subsequently required for manual proofreading, checking, and corrections will almost certainly increase. Single-pass keying may be sufficient when the text file is likely to need extensive editing at a later stage anyway. In other situations, double or triple keying may be more economical in the long run.

These considerations also apply when a single version of the text is chosen for conversion to electronic form, despite the existence of multiple printed or manuscript states of the text. Chadwyck-Healey's literary databases, for instance, generally only contain a "best text"— a single version of a printed edition. The version chosen is usually a first edition or one from the author's lifetime. If this is not sufficiently reliable, a later version may be used instead. Variations between editions are not normally recorded. The different printed editions of Migne's *Patrologia Latina*, for example, are not reflected in the *Patrologia Latina Database*. It uses only the first edition, published between 1844 and 1855, rather than the inferior reprint made by Garnier in the 1860s. Chadwyck-Healey's methodology is generally to have the full text keyed in twice, independently, and for the resulting files to be matched against each other for inconsistencies and variations. Further manual proofreading is then carried out.

When an electronic edition aims to include all the different versions of a text, one approach is to key in each version several times. The alternative is to key in each version only once, accompanied by rigorous checking and proofreading. The *Canterbury Tales Project*, which has as its goal the publication of all pre-1500 versions of Chaucer's poem, adopts the latter approach. In part, this is because of the sheer scale of the work involved, with as many as eighty-eight different versions of the text needing to be keyed in. But the highly specialized skill required to transcribe a Middle English manuscript is also a major factor. The single-keyed transcription is checked at least three times, with an error rate of no more than 1 in 4,000 characters expected on the third check. This is the same as an accuracy rate of 99.96 percent.

Clear Rules for Keying in the Text Must Be Established

The rules and instructions for keying in the text will differ, depending on the expertise of the person doing the keying and on the kind of markup being used for the initial file. But, rules for transcription will always be necessary, however straightforward the task appears to be. If, at the simplest level, a printed English-language text is being keyed by a bureau typist into a common word-processing file format, the instructions must insist on no corrections or alterations to the spelling and punctuation of the text. It must be typed exactly as it stands. Guidance must also be given on how to handle structural features (such as headings, lists, lineation, and pagination) and instances of formatting such as bold or italics. They may need to be indicated in some way or ignored altogether.

If the source being transcribed is more complex, it will almost certainly be necessary to employ an expert for the process. This is inevitable for manuscript materials, for which specialized knowledge of writing systems and languages other than English is usually required. It is also likely to be the case when the physical and structural features of the original document must be preserved in the electronic version. Detailed and elaborate rules for transcription must be developed and applied consistently and thoroughly. In many ways, these rules are analogous to those used in preparing editions of manuscripts for publication in printed form. Peter Robinson (1993c:10) sees "no division between transcription and editing. To transcribe a manuscript is to select, to amalgamate, to divide, to ignore, to highlight, to edit." The only differences arise from the nature and purpose of the markup being used for the electronic text.

One important aspect of these rules for transcription is what might be called the symbolic or the representational. This refers to the way in which markup is applied to represent the various features of a specific text. These rules for applying markup are necessary regardless of the kind of markup scheme being used. Whether the TEI scheme is being used, or a different—perhaps unique—approach, the transcriber must have clear instructions that relate the markup scheme to the distinctive characteristics of the text being encoded.

They are particularly important when the formal features of a manuscript are being recorded. How should an illuminated initial be represented in the electronic text, for instance? What about scribal punctuation? How are contemporary foliations to be handled? How should text in multiple columns be represented? The transcription system should be capable of answering these questions consistently, using a mixture of standardized markup and interpretative rules.

These rules about applying markup consistently to a text are only part of the story. There is another layer of policies and practices that apply to any kind of edition, not just to an electronic one. This might perhaps be called the intellectual aspect of transcription and involves more fundamental questions. Among the issues on which a ruling might be required are the following:

- Are the line divisions to be preserved?
- Is contemporary foliation or pagination to be recorded? What if there are multiple foliations?
- Should running headings be transcribed?
- Will a columnar format be recorded?
- Are later annotations or insertions to be transcribed or ignored?
- Will the orthography of the manuscript be preserved or normalized?
- Which takes precedence—preserving an abbreviation or expanding it?
- Which takes precedence—the original text or a subsequent correction or deletion?

- How should scribal punctuation be recorded?
- Should word separation be normalized or not?
- How should nontextual features (illuminations, diagrams, miniatures) be recorded?
- Should rubrics be recorded as such?

These, and many other such questions, will have to be answered for any kind of transcription, not just for an electronic version of the text. They form the intellectual heart of the process of transcription.

For any project involving manuscript materials, consistent rules for transcribing the text are essential. This is particularly so for larger-scale collaborative enterprises. Robinson (1994:12-28) discusses the transcription systems used for several major electronic text projects, among them the *Hartlib Papers*, the *Wittgenstein Archives*, and the *Dictionary of the Old Spanish Language*. For the most part, these use symbols such as brackets, braces, asterisks, and slashes to indicate features such as headings, layout, conjectural readings, insertions, and deletions. The *Manual of Manuscript Transcription for the Dictionary of the Old Spanish Language* is an excellent example of detailed transcription rules for medieval Spanish manuscripts (Mackenzie, 1986). It provides guidance on a wide range of subjects: foliation, headings, columns, characters, word separation, abbreviations, deletions, insertions, rubrics, initials, diagrams, glosses, and so on. The rules cover general questions of what to transcribe and how to transcribe it, as well as specific matters arising from the markup scheme used.

The best-documented transcription rules are probably those of the *Canterbury Tales Project* (Robinson and Solopova, 1993). Though they are intended only for the specific manuscripts covered by the project, and not as a standard approach for medieval English texts, they illustrate nevertheless some of the fundamental choices to be made in the transcription process. A regularized transcription was rejected for the project, on the grounds that this would be difficult and time-consuming and would actually lose important information. Regularization can be performed by the project's collation software, Collate, after the transcriptions are completed. At the other extreme, a "graphic" transcription, which would record every mark and space in each manuscript, was also rejected. Instead, the visual appearance of the text would be captured by facsimiles. A third, "graphemic" approach to transcription was adopted, which preserves every spelling in the manuscripts but not every distinct letter type. The transcription also aims to record all the symbols, abbreviations, and special characters used in the text, as well as its capitalization and punctuation.

A more generic basis for transcription may be found in the TEI *Guidelines*, which contains advice on the use of TEI-based markup in transcribing manuscripts (Sperberg-McQueen and Burnard, 1994: 529-557). The guidelines cover such topics as:

- abbreviation and expansion,
- corrections and conjectures,
- additions and deletions,
- substitutions,
- damaged and illegible text,
- empty space, and
- headers and footers.

These are the kind of features typically covered by transcription rules. But *Guidelines* also makes provision for indicating different scribes at work in the manuscript and for recording where a new scribe takes over. Although of considerable paleographical significance, this information is rarely found in transcription rules for electronic texts. Despite the level of detail involved, there are still some subjects not covered by TEI's *Guidelines*, as the editors note: the materials used, the layout, the organization of the physical text (quires, collation, and so on), and authorial instructions or scribal markup.

For all this, the TEI *Guidelines* are generic instructions, aimed at the most common characteristics of manuscripts. Applying them to a specific document will require interpretation and extrapolation. This will almost certainly mean a further set of rules for transcription in the context of a specific project. These rules may need to cover three areas: the general principles adopted for deciding which features should be transcribed, the markup guidelines to be followed, and the instructions and interpretations necessary for applying these markup guidelines in the specific case.

REUSING AN EXISTING TEXT

It may be possible to avoid the complications and details of keying in a text if it already exists in an electronic form. Most important, the existing electronic version must be available for reuse. Some electronic texts are available freely in the public domain, from archives such as the Oxford Text Archive and the *Online Book Initiative.* But others are restricted, and permission for any reuse must be sought from the owner of the electronic version. This is in addition to any permission that may be required from the owner of the copyright in the original text.

Reusing an existing text is worth doing if the aim is to add value to it in some way. Adding more sophisticated markup to a word-processed or "plain ASCII" file would be a suitable example of this process, as would converting an HTML file to TEI markup. A project of this kind is described by James Tauber, who took electronic texts of the Greek *New Testament* and Dante's *Divina Commedia,* from the University of Pennsylvania's Center for the Computer Analysis of Texts, with minimal markup, and transformed them by adding TEI encoding (Tauber, 1995). The Dante file was later made available as a searchable and browsable text through the Electronic Text Service of the University of Western Australia's Scholars' Centre (Burrows, 1996).

There are several important issues that must be investigated before an electronic text is reused in this way, to ensure the scholarly validity of the eventual result. Foremost among these is the provenance of the existing text. Which printed edition or manuscript was used to create the electronic version? This information must be documented in the text itself, or in its ancillary materials, if the electronic version is to be a credible and reliable tool for scholarship and research. For much-edited authors such as Dante and Shakespeare, identification of the source of the existing electronic text is particularly important. A text with origins in an unsatisfactory printed edition or in the eclectic choices of an idiosyncratic inputter will not stand up to scholarly scrutiny. At the very least, its origins should be clearly stated and documented, but it may be better left unused. Information about the origin of the text is doubly necessary if the eventual aim is to provide electronic versions of a range of different states of the text based on a variety of manuscripts or on several printed editions.

The accuracy of the transcription in the existing file is also extremely important. Some knowledge of the process by which it was created will be highly desirable. Who keyed it in? What measures for checking and verifying the accuracy were employed? If there is any doubt about these questions, it will be necessary to check the text against its printed or manuscript source. This is yet another reason for needing to know the source of the text.

Another important consideration is the transcription rules and practices followed in creating the existing electronic file. Important information may have been lost, inconsistencies may have crept in, and structural and formatting characteristics may not have been captured fully or in a usable format. The text may have been "corrected" or normalized to some degree.

Unless all these matters have been clearly documented, it may be better to begin again and rekey the text rather than reusing the existing file. At the very least, it will be necessary to check the file closely against the source, to identify corrections, remove inconsistencies, and add missing information.

The practices adopted toward this kind of documentation of an electronic text vary greatly between sites on the Internet. One of the criticisms often made of *Project Gutenberg*, which produces widely distributed "plain ASCII" texts, has been that the sources of its texts are not sufficiently documented. Most of the project's texts do not record their source, but the latest texts do now carry a brief statement of their provenance. The 1998 version of three ghost stories by Charles Dickens, for example, carries the statement, "This text was prepared from the 1894 Chapman and Hall 'Christmas Stories' edition" (Project Gutenberg, 1998). However, no indication is given of the practices and standards adopted by *Project Gutenberg* for transcribing texts. This information would be of considerable relevance, given that transcriptions are carried out by numerous volunteers.

The Oxford Text Archive, on the other hand, usually records the provenance of its texts in considerable detail. Since the archive serves as a clearinghouse for texts created by other scholars, the descriptions are mostly those provided by the original creator of the electronic version. But the information generally covers the printed edition from which the electronic text is derived, together with some record of the people involved in the creation and revision of the file. Transcription policies and practices are not usually documented, however.

Sometimes, electronic texts come with a warning to the reader of possible inadequacies in the text's provenance. The version of Dante's *Epistle to Cangrande,* available through the Center for the Computer Analysis of Texts at the University of Pennsylvania, carries the following engaging disclaimer:

> This text was supplied by James Marchand of the University of Illinois. It is meant only as a convenient reading text and should not be preferred to any available critical edition. Since the text was scanned from an old typescript, caveat lector, particularly for confusion of n, u, a and also of c, o, e.

SCANNING THE TEXT

If a text exists in printed or manuscript form, but not as an electronic file, another option worth considering is to scan it in from the printed original. This involves using either a digital camera or a scanner to take the digital equivalent of a photograph for storage as a computer file. The scanned digital image can be further manipulated and processed using OCR software to create a character-based text file.

Creating a Digital Image of the Text

A digital image is a computer reproduction of the visual appearance of any physical object or scene. Although we are only concerned here with the representation of pages of books or manuscripts, a similar process is applied to the digitization of photographs, motion pictures, and video. The computer reproduction consists of a series of dots known as pixels, or picture elements. Each pixel is derived from a value expressed in bits (0 or 1), that the computer interprets as an instruction to display a particular color.

There are two major variables that apply to digital images:

- The range of colors that can be expressed as the value for each pixel and displayed by the computer's monitor.

- The resolution at which the image is displayed, derived from the number of pixels in the image.

For the first of these, known as *bit depth* or dynamic range, there are three main categories: bitonal, grayscale, and color. A bitonal or binary image contains only black and white. Each pixel has a value of either 0 (for black) or 1 (for white). To indicate gray, a process called dithering may be used, which involves grouping black and white pixels in such a way as to give the appearance of gray. For showing different shades of gray, however, each pixel needs to be given a more detailed value. Instead of using only one bit to indicate the color, two bits can be used, to give a choice of four values: 11 for white, 10 for light gray, 01 for dark gray, and 00 for black. The greater the number of bits used for each pixel, the greater the number of shades of gray that can be indicated and displayed. An 8-bit pixel (00000000) can show any one of 256 different values, which can be used to distinguish 256 different shades of gray.

The same principles of this grayscale can be applied to colors. If each possible value for a pixel is applied to a different color, an 8-bit color image can contain up to 256 different colors. Also in increasing use are 24-bit images, in which each pixel has a value composed of 24 bits and can indicate one of more than 16.7 million different colors. In the RGB (red-green-blue) format, these are grouped into three 8-bit values, the first representing the redness of the pixel, the second representing its greenness, and the third its blueness. As if this was not enough, 36-bit image files are now beginning to appear as well.

In addition to this "bit depth," with its effect on the range of colors that can be displayed, image files are fundamentally affected by the number of pixels they contain. The standard measure of this is the *resolution* of the image, expressed in terms of "dots per inch," or dpi. An image rendered at 100 dpi contains 100 dots (pixels) per vertical or horizontal inch, or 10,000 dots per square inch (100×100). A resolution of 200 dpi means 40,000 dots per square inch. The resolution of a typical computer screen is around 70 to 100 dpi, while laser printers work at 300 dpi. In comparison, a typeset book has a resolution of about 1,200 dpi.

The storage space required for an image file is directly related to both its resolution and its bit depth, as well as to the surface area of the document being digitized (Robinson, 1993a:11-12). For a computer reproduction of an A4 page, stored as a bitonal image at a resolution of 100 dpi, the resulting image file will be about 112 kilobytes in size. A grayscale 8-bit image at the same resolution would be 900 kilobytes, while a color 24-bit image would be 2.7 megabytes. A color 24-bit image at a resolution of 400 dpi would result in a file of 43.2 megabytes. Such an image would be about as large as the electronic text of forty books, and only about fifteen of these images would fit on a CD-ROM.

Because digital images can quickly reach such a large size, there are various methods of compressing them using mathematically based algorithms. Some of these techniques are designed for bitonal files, others for grayscale or color images. The crucial factor is whether the compression is "lossless" or "lossy." Lossless compression means that the viewed image is an exact replica of the original image; no information is discarded during the process of compression. Lossy compression, on the other hand, means that some less significant information is discarded or averaged during the process of compression. Consequently, the viewed image is not an exact replica of the original image, even though it may appear identical to the human eye.

Various types of compression are available. For bitonal images, a widely used lossless technique is the CCITT/ITU (Group IV Fax) standard for interchange of fax images. This is particularly applicable to images of text. It can produce an average lossless compression of around 20:1 for text files (Kenney and Chapman, 1996:23). For grayscale and color images, there is the JPEG standard (ISO 10918-1) that is lossless at up to 3:1 but lossy at higher rates of compression. Also widely used is Kodak's proprietary ImagePac technique, which appears to be lossless but is in fact lossy, even at the highest resolution; Robinson (1993a:52) states that it "compares poorly with JPEG."

There are various different file types that can be used for digital images. Which type is applicable in a specific context will depend on a combination of factors: whether an image is bitonal or grayscale/color, the kind of compression used, the likely method of delivering the image to the eventual user, the degree to which the image needs to be transferable between systems, and the user's likely requirements for printing the image.

The following are a few of the more common file formats:

- TIFF (Tagged Image File Format): uncompressed grayscale/ color images; also used for bitonal images compressed using CCITT/ITU.
- Kodak PhotoCD (ImagePac): a proprietary format that provides five formats for each image, ranging from thumbnail size to a resolution of around 300 dpi. (See Robinson, 1993a:48-52.)
- GIF (Graphic Image File, or Graphics Interchange Format): originally promoted by CompuServe and widely used for images on the World Wide Web, with lossless compression and 8-bit depth.
- JFIF (JPEG File Interchange Format): used for images compressed with the JPEG algorithm, with lossy compression and either 8-bit or 24-bit depth.
- PDF: Adobe Systems' proprietary Portable Document Format, widely used for images of scholarly journal pages and viewable with the free Adobe Acrobat Reader software.
- PICT: a format designed primarily for Macintosh computers.
- PNG (Portable Network Graphics): a new format that has several technical improvements but is currently not supported by the Netscape Web-browser software. It has lossless compression and 8-bit or 24-bit depth.

Probably the most crucial aspect of choosing among different image formats is the combination of software and hardware with which the image will be stored and used. Some formats need far more specific and specialized platforms than others. Delivery over the World Wide Web depends, in part, on the Web-browser software being used. Netscape, for example, will display GIF and JPEG images, but requires additional "plug-in" viewer software to display PDF or TIFF files. Software for viewing and manipulating images is discussed in more detail in Chapter 3.

An important consideration with Web images is what is known as their palette. Not all computers are capable of displaying the full 16.7 million colors contained in 24-bit images. Those which can only handle 8-bit or 16-bit color must match, or dither, the colors to fit their more limited range. This finite selection of colors is the palette. To ensure that a color image will display well over the Web on machines with 8-bit color, it may be advisable to stick to the colors in the so-called "Web-safe" or "browser-safe" palette. This contains 216 colors, out of the 256 that are possible with 8-bit images. The other forty colors vary between Macintoshes and PCs. An image can be manipulated with software such as DeBabelizer to limit it to these "Web-safe" colors.

Another factor in creating Web images is the speed with which they will be loaded by browser software onto the user's screen. It is less than satisfactory to have to wait for some time, until the whole image has been transferred, before anything appears on the screen. This delay can be significantly reduced by a process known as interlacing, or progressive rendering. This involves storing the image in such a way that it loads one line at a time rather than as a whole. Most image manipulation software allows interlacing to be applied to images. For GIF files, only one method of interlacing is possible. For JPEGs, several options exist for choosing the order in which lines will load. PNG interlacing is done by pixels rather than lines and is claimed to be much faster than for other formats.

Scanning a text into an image file produces a digital picture of the text, but not a text that can be searched by keyword. Nor can the text be marked up for further analysis and manipulation. Nevertheless, an image file may still be a valuable way of representing a text. Above all, this process captures the visual appearance and layout of the text—its use of colors, its diagrams and illustrations, its physical structure, its mise-en-page. With the appropriate software, the image can be blown up to reveal details that are perhaps difficult to distinguish on the original. For manuscripts, in particular, it is well-nigh essential to accompany the electronic text with a facsimile in the form of a digital image.

The major advantages and disadvantages of this kind of digitization are well summarized by Robinson (1993a:12-14). He identifies the following disadvantages:

- Image files tend to be large and are consequently expensive to store.

- Manipulation of large images still tends to require a larger machine than a personal computer because of the amount of memory required.
- Compressing and decompressing image files can be time-consuming and awkward.
- The different formats for image files are not usually compatible with each other or with a range of different delivery systems.
- Image files can only be used through a series of underlying technical systems, which are complex and subject to rapid change. The files must be migrated to the latest systems to avoid being unusable because of obsolete hardware or software.

Robinson also identifies a number of advantages in creating digital images of texts:

- Image files will not decay in quality over time, unlike most films and microfilms, as long as the appropriate migration between systems is undertaken.
- Each copy of a digital image is perfect and does not affect the quality of the original file.
- Similarly, an image file is not degraded as a result of use—unlike slides and microfilms.
- Databases of image files, if suitably designed, are far easier to navigate to find a specific image than, for example, a microform set is.
- Digital images can be distributed globally over computer networks relatively easily and can be used by many people apparently simultaneously.
- Image files can be manipulated in various ways, such as altering their scale and placing them side by side on a screen.
- The cost of distributing digital image files on CD-ROM is much less than the cost of publishing a printed facsimile or even a microform.

From the point of view of the scholarly user or researcher, these advantages far outweigh the disadvantages. For a person or organization involved in creating, disseminating, and preserving digital images, an additional factor to be considered is the type and cost of the equipment needed to produce such images. Scanners and digital cameras range from comparatively inexpensive handheld devices to very costly freestanding, high-resolution apparatus such as the Kontron camera. The cost per image will increase in proportion to the power and sophistication of the equipment used. It is likely to be preferable, in all but the largest institutions, to contract out this digitization to a commercial bureau that has the appropriate range of equipment.

Whether the work is contracted out or not, very careful attention must be paid to documenting the standards of quality required. It will be necessary to specify the type of image required, the type of file to be used, the resolution needed, and the method of compression to be employed. In the case of large projects, there are likely to be different strategies for each category of material being digitized, depending on its specific requirements for quality. The different uses that will be made of an image—such as archival preservation or browsing by users—will also need different strategies for digitization.

The Library of Congress, for its *American Memory* project, has identified at least six different types of images for varying categories of material and uses (Fleischhauer, 1996):

- Grayscale/color images at 150 dpi, using JPEG compression in a JFIF format, for reference copies (i.e., those displayed by a user).
- Grayscale/color images at 300 dpi in uncompressed TIFF format, for the archival copy.
- Grayscale/color images at 300 dpi, using JPEG compression in a JFIF format, for compressed archival copies.
- Bitonal images at 150 to 300 dpi, using CCITT/ITU compression in a TIFF format, for reference copies.
- Bitonal images at 300 to 1200 dpi, using CCITT/ITU compression in a TIFF format, for compressed archival copies.
- GIF images at 75 to 100 dpi are envisaged for Web browsing.

The Electronic Text Center at the University of Virginia Library recommends 24-bit color at 400 dpi as the default setting for scanning from material in their Special Collections (Seaman, 1996a). A resolution of 600 dpi would be preferable, but limitations on storage capacity impose the lower limit. Even a resolution of 400 dpi results in comparatively large TIFF files, which are stored on CD-ROMs rather than a server. From these, one or more JPEG versions of the image are created for on-line use. All images, including grayscale book illustrations and engravings, are produced by 24-bit color scanning. This methodology results in smaller, and often better, JPEG files than is possible with 8-bit color or grayscale scanning.

The *Duke Papyrus* project follows the 600-dpi standard adopted by the Advanced Papyrological Information System (APIS), after having previously used a resolution of 300 dpi. This standard allows images to be enlarged eightfold on most computer screens, in line with the enlargements of between four and ten times that papyrologists commonly use for studying documents under a microscope. Several different formats are stored for each image. The archival master is a 600-dpi TIFF file, and a JPEG version is also made. These are available to researchers within Duke University. For Web browsers, two lower-resolution scans are provided, at 150 dpi and 72 dpi.

Kenney and Chapman (1996:33) summarize scanning guidelines from a range of North American imaging projects:

- Published text: bitonal images at 600 dpi
- Illustrated text: bitonal images at 600 dpi, or 8-bit grayscale at 300 to 400 dpi, with 24-bit color scanning for color illustrations
- Archival documents: bitonal images at a minimum of 300 dpi for typewritten or laser-printed documents, and for most ballpoint; bitonal or gray at 300 dpi for pencil, quill, or felt tip; 24-bit color at 600 dpi for papyri

They also recommend the adaptation of micrographics standards to assess the quality of digital images, using the ANSI Quality Index to relate image resolution and the quality of text. It is based on the "x-height," the height of the smallest significant text character, measured in millimeters. The digital Quality Index developed by Kenney and Chapman (1996:17-19) takes the form:

$$QI = (.039 \text{dpi} \times h)/3$$

for bitonal images, where h is the x-height. For uncompressed grayscale and color, they suggest a slightly different formula:

$$QI = (.039 \text{dpi} \times h)/1.5$$

From these, it is possible to derive recommendations for resolution requirements for marginal-, medium-, and high-quality images at a range of x-heights (Kenney and Chapman 1996:24-25).

Performing Further Processing
on the Digital Image (OCR)

An image file can be processed further, using Optical Character Recognition (OCR) software. This takes a file in any of the more common formats—compressed or uncompressed—and creates from it a searchable electronic version of the text involved. In most cases, this kind of software is capable of creating the original image file as well as producing a subsequent text file. However, it is also possible to capture the image file with a different method and then apply OCR to it separately. The image file is not deleted by OCR processing.

OCR programs follow four main steps to produce a text file:

- Rotating the image to its correct orientation: there is considerable variation in the extent to which slightly skewed images can be processed effectively.
- Analyzing the layout and structure of the page: this is especially important where multiple columns are involved, as well as for tables, graphics, captions, and footnotes.
- Identifying individual characters: algorithms are used to match clusters of pixels to stored patterns for letters of the alphabet, numbers, and punctuation marks. This matching can be done against templates from font files or by recognizing features such as joins and curves as components of characters.
- Applying contextual information: this focuses particularly on common patterns of word usage and grammatical structures.

Many OCR software packages are available. One of the more widely used is OmniPage Pro. Its makers claim that it is "the world's most accurate" OCR software, with over 99 percent accuracy. By this, they mean that the average error rate is below 4 words out of every 400, for documents of laser-printed quality that use standard fonts. Among the tools for recognizing characters is the Language Analyst, which draws on a dictionary and linguistic information for its analysis of a text. The scanned image can be compared on the screen with the OCR version, and the software will highlight characters that are doubtful and will suggest corrections, similar to a spell-checker program. OmniPage Pro can be used from within other software such as Microsoft Word and Excel. It will also convert documents to HTML markup.

Another example of OCR software is TextBridge Pro, developed by Xerox. It works with a wide range of scanners and can process files in several different image formats. It can also be integrated with Word or WordPerfect in various ways, including the ability to proofread a scanned document from within the word-processing program. Conversion of files to HTML is also offered. The accuracy with which documents are analyzed is derived from two proprietary features:

- Dynamic Training—enables the software to be trained to recognize particular symbols and words whenever they occur.
- Lexical Technology—enables the software to apply recognition techniques based on words or other valid patterns (such as dates), in addition to the usual character-by-character method of recognition.

The effectiveness of OCR is heavily dependent on the nature of the original document being scanned and analyzed. The best results are likely to be achieved with documents that have high contrast, regular spacing, and a single font and which have been scanned at a resolution of 300 to 400 dpi. At present, bitonal images are handled best, although OCR for grayscale images is expected to improve rapidly in the near future. OCR does not work effectively for documents that contain small type sizes, older or ornamental fonts, or a mixture of Roman and non-Roman scripts. Nor is it usable for color images or handwritten originals. A negative effect will also result from the use of lossy compression.

The resulting accuracy rates may be quite low. Even an achieved rate of 95 percent results in approximately 100 errors on a page of 2,000 characters. It may be possible to maximize the accuracy by passing the same image through several different OCR programs in succession, but 100 percent accuracy is unlikely. This has a major effect on cost. Each error must be identified and corrected manually, at a considerable cumulative cost, which may quickly become more expensive than direct input. One study has concluded that OCR is less efficient than manual keying when accuracy rates fall below 95 percent (Kenney and Chapman, 1996:135). For the kind of material involved in humanities scholarship, even achieving a rate of 95 percent is unlikely.

Given these limitations and potential costs, OCR may have a somewhat restricted applicability in the creation of scholarly texts in the humanities. The original documents are too heterogeneous, and the need for complete accuracy is too pressing. Nevertheless, there may be circumstances in which the source material is such that a reasonable result can be expected from the use of OCR. This is especially likely when a printed version of a text is being scanned, and the quality and style of the printing are sufficiently clear and large. Using OCR will save considerably on the time necessary to key the text in, especially if double-pass keying is required. In fact, the speed of OCR software may be up to 100 times faster than manual keying, depending on the equipment being used.

The time-consuming part of the OCR process is the subsequent reviewing of the text. Careful proofreading and checking will be absolutely essential and can take a substantial amount of time. The relative merits of the two processes—OCR and manual keying—will need to be weighed, perhaps through applying both of them to a sample of the text and then comparing the results. If encoding with SGML-based markup is considered an essential characteristic of the electronic version of the text, this should be added at a stage subsequent to the OCR process. When this kind of markup can be added during the keying of the text, the manual process may prove to be more efficient. Despite all the obvious limitations, certainly, a role exists for OCR in the creation of electronic texts for scholarly purposes. However, this role needs to be assessed and evaluated carefully and applied when the situation is right.

FURTHER READING

Peter Robinson has produced definitive guides to digitization (Robinson, 1993a) and transcription (Robinson, 1994), which are aimed specifically at scholars in the humanities. The essays by Popham and Robinson in the collection edited by Mullings et al. (1996) both provide good introductions to text creation and image capture, respectively.

A detailed and very helpful practical guide to the creation of digital images was compiled at Cornell University by Anne Kenney and Stephen Chapman (1996). A shorter introduction to the same field is provided by Howard Besser and Jennifer Trant (1995), and is also available at the Web site of the Getty Information Institute.

Information about image formats can be found through the CICA Graphics List Web site. A useful short introduction is given by Fleishman (1997). The browser-safe palette can be seen at Lynda Weinman's Web site.

There are useful basic help sheets on text and image scanning at the Web site of the University of Virginia's Electronic Text Center.

A technical introduction to OCR and related matters can be found in O'Gorman and Kasturi (1995).

WEB SITES

Web sites related to HTML and SGML software are listed in Chapter 3.

Canterbury Tales Project:
http://www.shef.ac.uk/uni/projects/ctp/index.html

CICA Graphics List—Image File Formats:
http://cica.indiana.edu/graphics/image.formats.html

Dante's *Epistle to Cangrande* (the text of) can be found at:
gopher://ccat.sas.upenn.edu:70/00/journals/Recentiores/Dante/Cangrande.Latin

Duke Papyrus project: http://odyssey.lib.duke/edu/papyrus/

Getty Information Institute (Besser and Trant, *Introduction to Imaging*): http://www.ahip.getty.edu/intro_imaging/0-Cover.html

Lynda Weinman, The browser safe color palette:
http://www2.lynda.com/hex.html

OmniPage Pro:
http://www.caere.com/live/content/products/amaretto/amaretto.htm

TextBridge Pro: http://www.xerox.com/

University of Virginia, Electronic Text Center:
http://etext.lib.virginia.edu/

Chapter 3

Delivery Mechanisms for Electronic Texts

After an electronic text has been created, it must be published and distributed. Although it is perfectly possible to leave the file on one's own computer, this is equivalent to having a typescript in the desk drawer; no one else is likely to see it or use it. For an electronic text to be seen and used by other people, there must be a delivery mechanism of some sort. In this chapter, we look at the two major aspects of such delivery mechanisms: the media that are used to transmit electronic texts and the software by which texts can be used and analyzed.

Transmission media perform the function of getting the text to its user. This can be done in a variety of ways: on diskette, on CD-ROM, on magnetic tape, and across the Internet and the World Wide Web. Once a transmission medium has been chosen, appropriate software must be available for the user to read and analyze the text. This software may be generic word-processing or viewing software. In addition, numerous specialized kinds of software exist that are designed for handling electronic texts, particularly those in an SGML-based format.

TRANSMISSION MEDIA

Diskettes

In earlier years, electronic texts were likely to be distributed on diskettes. A good example of this approach, published in 1990, is a version of the works of David Hume on eight floppy disks (Hume,

1990). These are low-density, 5¼-inch disks suitable for the IBM-XT type of personal computer. The files are "plain ASCII," intended to be used with word-processing programs. In their original form, they are no longer usable, except for situations in which a specific machine of this kind has been preserved in working order. The series of philosophical texts known as *Past Masters*, published by Intelex, were originally issued on diskettes in the late 1980s. The second edition, published in 1990, could be obtained on either 5¼-inch or 3½-inch disks; they ran on IBM-type personal computers with the DOS 2.0 operating system.

Texts such as these on diskette were often specific in the kind of operating systems they used, the type of file formats they employed, and the type of computer on which they would run. Given the rapid pace of change in computer technology in the past five to ten years, often resulting in the obsolescence and incompatibility of older equipment and software, this kind of electronic text has not proved to be a long-term proposition.

Apart from anything else, diskettes have too many disadvantages as a publishing medium. They have only a small capacity, perhaps less than two megabytes. Their format is likely to be a source of difficulties. Newer computers generally do not accept 5¼-inch disks, for example, and Macintosh and PC disks are not interchangeable without special software. To add yet further difficulties, disks created with specific operating systems—whether Macintosh or PC—may well be unusable on computers that have earlier or later versions of the same operating system. Windows 95, in particular, has created considerable problems of compatibility between newer and older PCs. To ensure that a text on diskette remains usable will require its reformatting into new versions of the operating system and probably also its transfer to newer standard diskette formats.

CD-ROMs

In practice, diskettes were quickly superseded by CD-ROMs, which have been used to distribute electronic texts since the late 1980s, and are still used extensively for commercially produced texts. In 1996 and 1997, for instance, Cambridge University Press has published several major literary texts on CD-ROM. These include Samuel Johnson's *Dictionary of the English Language* (the first and third editions) and the *Collected Works of John Ruskin* (originally published in thirty-nine large printed volumes between 1903 and 1912), as well as the first installment of Peter Robinson's remarkable edition of *The Canterbury Tales*—the *Prologue* to *The Wife of Bath's Tale*.

Chadwyck-Healey Ltd. publishes a wide range of texts on CD-ROM, including *English Poetry, Goethes Werke*, and the *Patrologia Latina Database*, and Intelex now distributes its *Past Masters* philosophy texts on CD-ROM. CD-ROMs have a considerable storage capacity—equivalent to about 450 diskettes or 300,000 printed pages. The entire second edition of the *Oxford English Dictionary* fits on a single CD, as does the *Encyclopaedia Britannica*. However, large compilations—such as most of the Chadwyck-Healey titles—require several disks.

Using a CD-ROM requires a personal computer with a suitable CD-ROM drive. Although these are now a standard component of most new computers, they must be bought separately and attached to older machines, along with the appropriate additional software for the operating system. Many older computers lack the capacity to run this software. Another important consideration is the speed of the CD-ROM drive itself. As CD-ROMs have become more sophisticated, especially with the inclusion of files in a variety of media (sound, images, and so on), faster drives are needed to run them effectively. Twelve-speed, and even twenty-speed, drives are now commonplace, but many libraries are battling on with quad-speed drives from only three years ago.

The operating system of the local machine is another major issue. Some CD-ROM texts will only run on a Macintosh, while others require older versions of Windows or even DOS. CD-ROMs are rarely designed to run under both the Macintosh and Windows operating systems, though this is perhaps becoming less unusual than it used to be. The problems arising from obsolete software and equipment are somewhat less acute than with diskettes, but, nevertheless, they do still occur and do affect the continued effectiveness of texts issued in CD-ROM format. In particular, many CD-ROMs now require Windows 95 software and cannot be run successfully on machines with other operating systems.

Networking such texts may also be a problem. Although it is possible to make CD-ROMs available over a local network, this really only works successfully where the network is a centrally controlled one and consists of the same generic type of machines running a single operating system, usually on a fairly limited scale. An internal network run by a university library is perhaps the best example of this type of arrangement. To network CD-ROMs across a whole campus is far more difficult, given the likelihood that this will involve machines running a variety of operating systems. Although software designed to operate such heterogenous networks is improving greatly, it is still very difficult to provide Macintosh users with access to Windows-based CD-ROMs, and vice-versa.

Even the size and capacity of CD-ROMs have proved insufficient for distributing electronic texts in the humanities. Although most publishers seem to try to tailor their product to fit on to a single CD-ROM, for obvious reasons, there are many texts that are simply too large and too ambitious. Chadwyck-Healey's *English Poetry* database, for instance, occupies no fewer than four CD-ROMs. To use them effectively requires a multiple CD-ROM drive, known as a jukebox, which only increases the potential for difficulties with obsolete or incompatible equipment and software. Without a jukebox, disks must be continually swapped in and out of a single CD-ROM drive, causing considerable inconvenience and annoyance to users, especially less experienced ones.

Magnetic Tape

A possible solution to some of these problems is to distribute electronic texts on magnetic tape. This method has been extensively used for publishing large bibliographic databases and indexes, but its use for electronic texts has been comparatively limited. It requires a large computer, known as a server, to which the texts on magnetic tape can be loaded for access through a local network, usually a campuswide one. A substantial degree of local computing expertise is necessary. The operating system for such a service is normally a variety of UNIX, and various kinds of software can be used to access the texts.

The best approach is probably to use what has been dubbed an "Intranet"—a kind of internal Internet that relies on Web-browsing software such as Netscape or Internet Explorer for its interface. Most larger North American universities have a networked electronic text service of this kind, but few libraries on other continents have followed this lead. In Australia, for instance, only the University of Sydney and the University of Western Australia currently offer such a service.

The Internet and the World Wide Web

The optimal method of distributing electronic texts, however, is to use the Internet. This approach has been around for quite a few years, but only began to become widely used in the early 1990s. By the Internet, I mean the decentralized physical network of cabling and equipment that now connects millions of computers around the world. These machines can communicate with one another because they understand various generic methods of communication, known as procotols. While the older protocols require the user on one computer to log in to another computer and to stay continuously connected for as long as the session lasts, the newer ones use a different method, known as client-server architecture. This involves a discontinuous connection; the client machine sends a request to a remote server asking for the transmission of a particular file or object, the remote server responds, and the connection is then broken. The client machine can then send another request to the same server or to a different server altogether.

The earliest method of distribution over the Internet was through the File Transfer Protocol, or FTP. This involved connecting to a remote computer, browsing its list of available files, and copying one or more files to the local computer, where they could be read with word-processing software. FTP was then joined, but not entirely superseded, by a method called Gopher, which used the client-server method for transferring text files. Both of these approaches are still used to some extent. Most of the collections of ASCII texts—such as *Project Gutenberg* and the *Online Book Initiative*—rely on FTP for the distribution of their texts. Even some SGML-based collections, such as the Oxford Text Archive, use FTP. The Center for Computer Analysis of Texts, at the University of Pennsylvania, still makes its extensive list of electronic texts available through a Gopher server; the texts themselves are also distributed by Gopher rather than by FTP. The *Wiretap Online Library* is another Gopher service with a collection of electronic texts. The *Dartmouth Dante* database is a Gopher service, as is the collection of several hundred ASCII texts provided by Virginia Tech (see, for example, Figure 3.1).

Both FTP and Gopher have now been incorporated into the World Wide Web and are usually accessed through it. The Web itself is based on the HyperText Transfer Protocol (HTTP), another client-server approach, but a far more powerful and versatile one than Gopher because of its support for multimedia files and its ability to embed links to other files within the files themselves. In little more than three years, the Web has become easily the most popular method for distributing noncommercial electronic texts. There are literally thousands of texts available through the Web, both personal collections such as James O'Donnell's and institutional collections such as those of the University of Virginia Library. Many of these texts are in HTML format, but the Web does not necessarily mean that texts must be stored as HTML. The Virginia corpus of Middle English texts, for example, stores texts in an SGML format and translates them automatically into an HTML display.

Commercial publishers are also beginning to move toward direct publication and distribution over the Web. At the end of 1996, Chadwyck-Healey Ltd. launched *Literature Online*, advertised as "the home of English and American literature on the World Wide Web." *Literature Online* offers thousands of electronic texts of poetry, drama, and fiction in English. A parallel service from the same company offers the *Patrologia Latina Database* over the Web (see Figure 3.2). Access to both these services is by payment of an annual subscription rather than by an outright purchase. This is a clear endorsement of the Web as a delivery mechanism. Although Chadwyck-Healey continues to distribute texts on CD-ROM for the time being, the emergence of Web services is likely to mark the beginning of the demise of CD-ROMs as a means for publishing large electronic texts.

There are major advantages in Web-based services, at least for the local user. There is no need for local storage or maintenance of large files or specialized programs. As long as preexisting local

FIGURE 3.1. Center for the Computer Analysis of Texts: Gopher Server (Partial List)

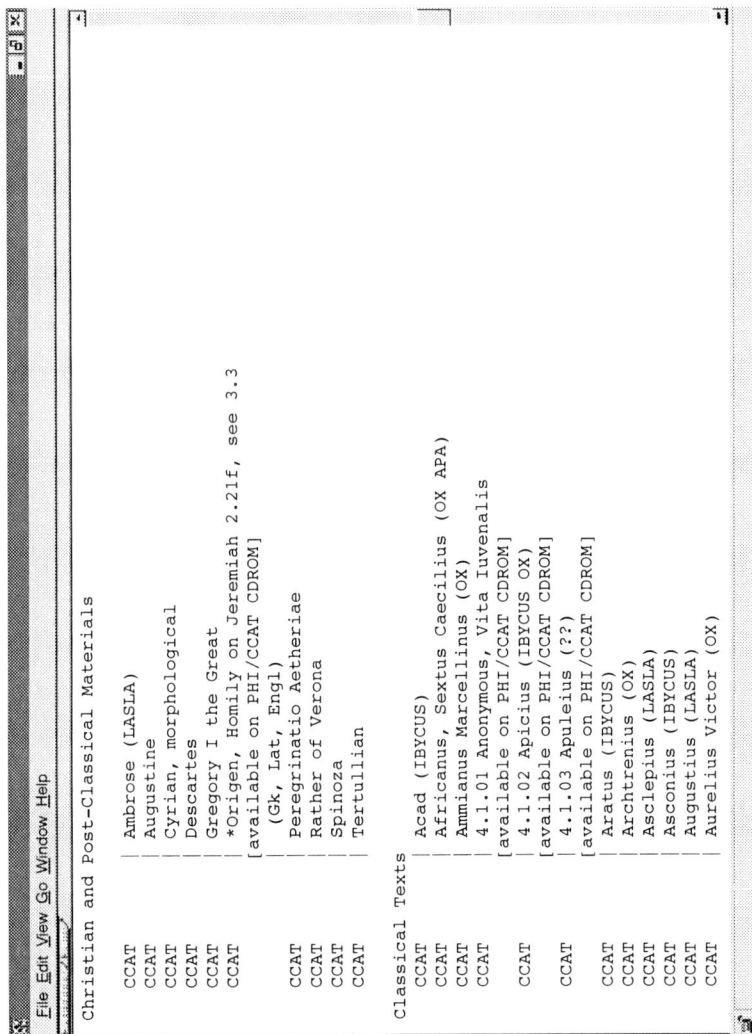

```
File  Edit  View  Go  Window  Help

Christian and Post-Classical Materials

     CCAT        | Ambrose (LASLA)
     CCAT        | Augustine
     CCAT        | Cyrian, morphological
     CCAT        | Descartes
     CCAT        | Gregory I the Great
     CCAT        | *Origen, Homily on Jeremiah 2.21f, see 3.3
                 [available on PHI/CCAT CDROM]
                 | (Gk, Lat, Engl)
     CCAT        | Peregrinatio Aetheriae
     CCAT        | Rather of Verona
     CCAT        | Spinoza
     CCAT        | Tertullian

Classical Texts
     CCAT        | Acad (IBYCUS)
     CCAT        | Africanus, Sextus Caecilius (OX APA)
     CCAT        | Ammianus Marcellinus (OX)
     CCAT        | 4.1.01 Anonymous, Vita Iuvenalis
                 [available on PHI/CCAT CDROM]
     CCAT        | 4.1.02 Apicius (IBYCUS OX)
                 [available on PHI/CCAT CDROM]
     CCAT        | 4.1.03 Apuleius (??)
                 [available on PHI/CCAT CDROM]
     CCAT        | Aratus (IBYCUS)
     CCAT        | Archtrenius (OX)
     CCAT        | Asclepius (LASLA)
     CCAT        | Asconius (IBYCUS)
     CCAT        | Augustius (LASLA)
     CCAT        | Aurelius Victor (OX)
```

FIGURE 3.2. Chadwyck-Healey Ltd.: *Patrologia Latina Database* (Home Page)

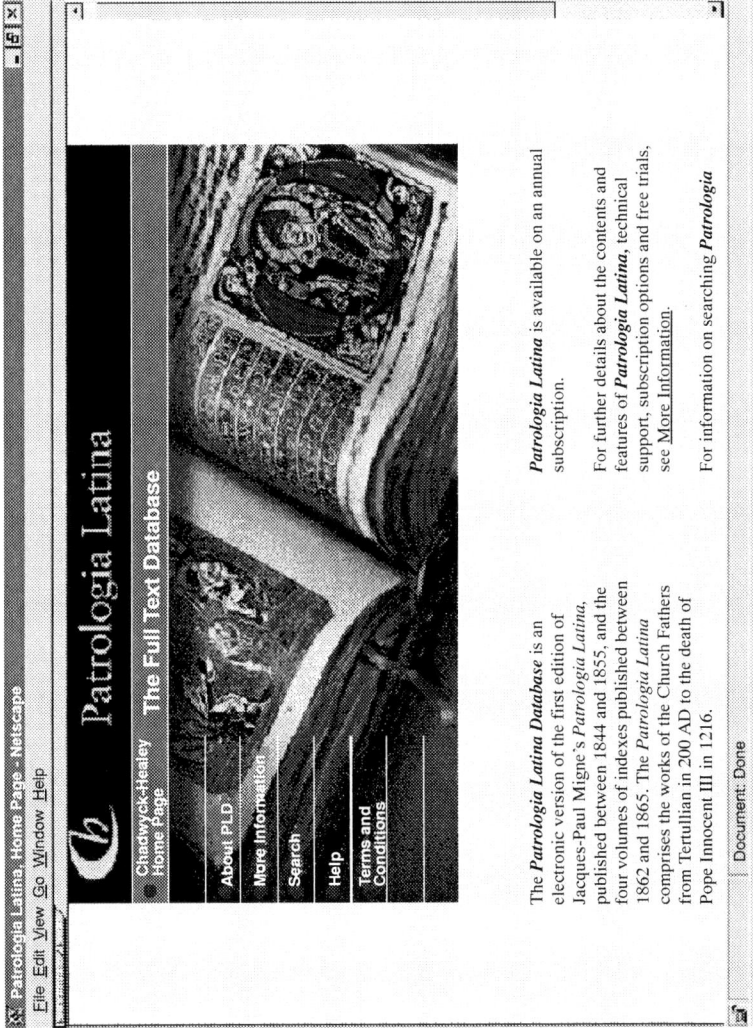

Patrologia Latina, Home Page - Netscape

File Edit View Go Window Help

Patrologia Latina

Chadwyck-Healey Home Page

The Full Text Database

About PLD

More Information

Search

Help

Terms and Conditions

The *Patrologia Latina Database* is an electronic version of the first edition of Jacques-Paul Migne's *Patrologia Latina*, published between 1844 and 1855, and the four volumes of indexes published between 1862 and 1865. The *Patrologia Latina* comprises the works of the Church Fathers from Tertullian in 200 AD to the death of Pope Innocent III in 1216.

Patrologia Latina is available on an annual subscription.

For further details about the contents and features of *Patrologia Latina*, technical support, subscription options and free trials, see More Information.

For information on searching *Patrologia*

Document: Done

network structures can be employed, no additional local computing expertise is necessary. Where desktop workstations are insufficiently powerful to run the latest Web-browsing software, however, promoting Web-based services may be something of a failure. Such services rely heavily on the speed and quality of Internet access, and they will be adversely affected by inadequate bandwidth, congested channels, and servers with too little capacity. Because they are dependent on generic Web-browsing software for their use, this is likely to result in a lower level of sophistication in searching and browsing than is possible with more specialized software.

SOFTWARE

Without software to interpret and display it, an electronic text is useless. This does not necessarily mean, though, that the same software employed in creating the text must be used to distribute and view it. Nor does it mean that the text should be designed for a single type of software. The most flexible and versatile electronic texts, in fact, are those which will work with various different types of software. In computer jargon, this is described as a high level of portability between software platforms. Publishing a text that requires a specific brand of software to read it and use it might be likened to publishing a book that can only be read with prescription spectacles from a specific firm of opticians!

Word Processors

An obvious approach to distributing electronic texts is to use the most commonly available software on personal computers: word-processing packages such as Microsoft Word and Corel's WordPerfect. The familiarity and ubiquity of this software make it the easiest way of creating and viewing electronic texts. There are many available varieties of word processors, most of which offer a wide range of helpful formatting devices. They work best with texts that are in their own proprietary format or in basic ASCII, but some are able to handle files in formats created by their competitors. Some brands of word-processing software also have the ability to read and produce files in HTML format, either intrinsically in the standard version of the software or as a separate product that may be added if required.

Despite its easy availability, word-processing software—at least on its own—suffers from the same kinds of disadvantages as specific operating systems do. Distributing a text in Word, for instance, raises questions of incompatibility between various versions of this software, especially those designed for different operating systems, let alone incompatibility with other kinds of software. The portability of such a text is somewhat low, unless its potential users have access to methods of converting files into other versions or other software. Another drawback is the limited extent to which this software can be used to search and manipulate the text. Although word-processing software of some kind may well be used in the creation of many electronic texts, it is one of the least satisfactory means for distributing and publishing them.

Similar observations apply to what is known as text-editing software, a term that normally refers to the equivalent of word processors for UNIX machines. The best-known text editors of this kind are pico, vi, and Emacs. In some cases, these are also available for Windows and DOS machines. Although they are generally more sophisticated than ordinary word-processing software in creating and manipulating a text, they are far more difficult to use than products with a so-called "graphical user interface"—no pull-down menus and automatic formatting here! Above all, however, these text editors are not designed to be a delivery mechanism. Although they can be used to open and read files, this is not a particularly easy or pleasant experience. They have the advantage of producing plain ASCII texts without proprietary markup, but there are far easier ways of reading the resulting files (see Figure 3.3).

Web-Browsing Software

If word-processing software is not used for distributing and viewing electronic texts, what alternatives are there? There are three main answers to this question. The first involves Web-browser software and HTML authoring software. The second is concerned with software for delivering image files, while the third encompasses a wide variety of software designed specifically for the creation, analysis, and publication of electronic texts.

CRT - Tobruk

File Edit Preferences Transfer Help

```
<l>vestite gi&agrave; de' raggi del pianeta</l>
<l>che mena dritto altrui per ogne calle.</l>
</lg><lg type=terzina>
<l>Allor fu la paura un poco queta</l>
<l>che nel lago del cor m'era durata</l>
<l>la notte ch'i' passai con tanta pieta.</l>
</lg><lg type=terzina>
<l>E come quei che con lena affannata</l>
<l>uscito fuor del pelago a la riva</l>
<l>si volge a l'acqua perigliosa e guata,</l>
</lg><lg type=terzina>
<l>cos&igrave; l'animo mio, ch'ancor fuggiva,</l>
<l>si volse a retro a rimirar lo passo </l>
<l>che non lasci&ograve; gi&agrave; mai persona viva.</l>
</lg><lg type=terzina>
<l>Poi ch'&egrave;i posato un poco il corpo lasso,</l>
<l>ripresi via per la piaggia diserta,</l>
<l>s&igrave; che 'l pi&ugrave; fermo sempre era 'l pi&ugrave; basso.</l>
</lg><lg type=terzina>
<l>Ed ecco, quasi al cominciar de l'erta,</l>
<l>una lonza leggera e presta molto,</l>
<l>che di pel macolato era coverta;</l>
</lg><lg type=terzina>
<l>e non mi si partia dinanzi al volto,</l>
<l>anzi 'mpediva tanto il mio cammino,</l>
<l>ch'i' fui per ritornar pi&ugrave; volte v&ograve;lto.</l>
</lg><lg type=terzina>
<l>Temp'era dal principio del mattino,</l>
<l>e 'l sol montava 'n s&ugrave;; con quelle stelle</l>
<l>ch'eran con lui quando l'amor divino</l>
</lg><lg type=terzina>
<l>mosse di prima quelle cose belle;</l>
<l>s&igrave; ch'a bene sperar m'era cagione</l>
<l>di quella fiera a la gaetta pelle</l>
</lg><lg type=terzina>
<l>l'ora del tempo e la dolce stagione;</l>
<l>ma non s&igrave; che, che paura non mi desse</l>
<l>la vista che m'apparve d'un leone.</l>
</lg><lg type=terzina>
<l>Questi parea che contra me venisse</l>
```

Ready   111 Cols, 41 Rows VT220   NUM

Web-browsing software is designed specifically for viewing files on the World Wide Web. The best-known and most widely used of these Web browsers are Netscape and Microsoft's Internet Explorer, which are available free or bundled in with other common software packages. Both offer graphical, "point-and-click" interfaces with a range of special features attached. But there are also various alternative browsers. Among these is Mosaic, the original graphic Web browser and precursor of Netscape, which is free but no longer being developed. Opera is a new Windows browser that takes up much less disk space than the two giants but offers most of their features. (See Figure 3.4 for an example of Netscape viewing.)

This type of software is designed to navigate and display files made available through the HyperText Transfer Protocol, and forms the "client" part of the client-server method of communication. Web browsers are the most common means of viewing electronic texts in HTML or ASCII formats. They also provide access to other types of files through what are known as helpers and plug-ins—other types of viewing software that are called on by the browser when it is presented with a particular type of file.

There are other kinds of software for browsing texts over the Internet, but these are of minor interest in comparison with Netscape and Internet Explorer. Among them is an SGML viewer called Panorama, which enables direct browsing of SGML, rather than HTML, files. Available for Windows, Macintosh and UNIX machines, the Panorama Viewer plugs in to Netscape and uses style sheets and navigation techniques designed to display full SGML markup. As recently as 1993, various kinds of Gopher browsing software were very popular, but these have been made redundant by the Web browsers and are now largely obsolete. (Remember Turbo-Gopher for Macintosh?) Also worth mentioning is the Web browser known as Lynx, which resembles the Gopher browsers in that it can only handle text (see Figure 3.5). The images, graphics, frames, and other programmed trickery so characteristic of graphical browsers such as Netscape are ignored by Lynx, which sticks resolutely to the text of HTML files. Lynx can handle forms, however, and is still a worthwhile option for users whose access to the World Wide Web is through terminals or slow modems, rather than personal computers with high-speed network connections.

FIGURE 3.4. Web Page Viewed with Netscape 4.0 (University of Western Australia Library)

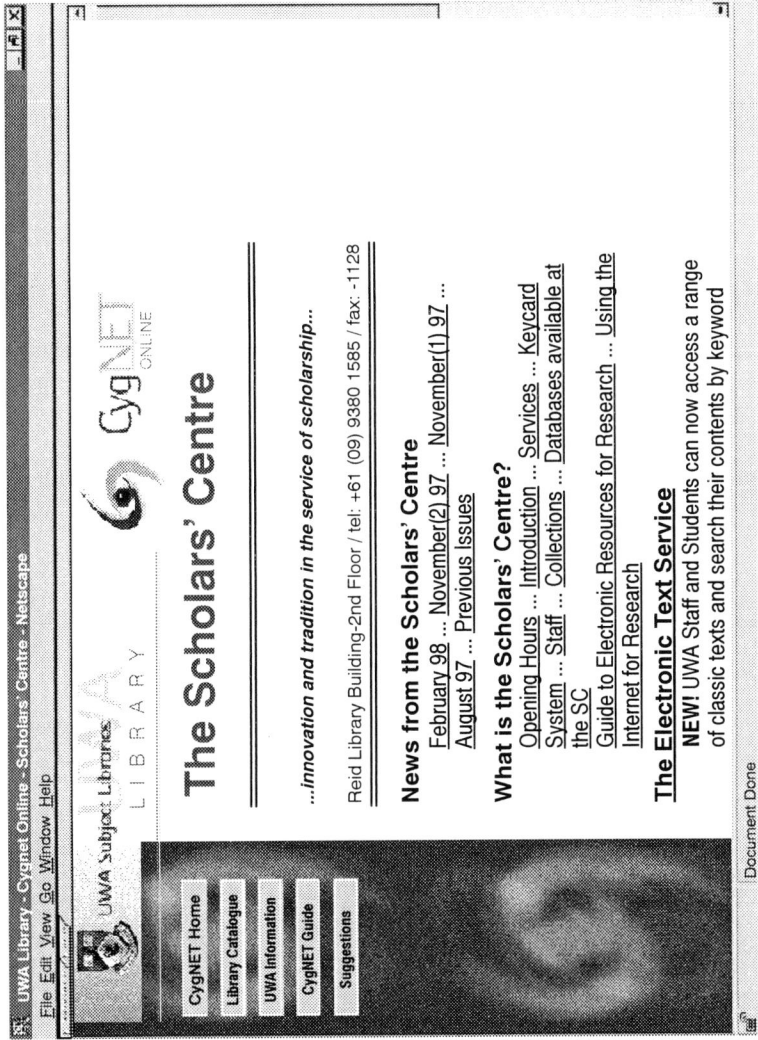

UWA Library - Cygnet Online - Scholars' Centre - Netscape

File Edit View Go Window Help

UWA Subject Libraries

LIBRARY

CygNET ONLINE

CygNET Home
Library Catalogue
UWA Information
CygNET Guide
Suggestions

The Scholars' Centre

...innovation and tradition in the service of scholarship...

Reid Library Building-2nd Floor / tel: +61 (09) 9380 1585 / fax: -1128

News from the Scholars' Centre
February 98 ... November(2) 97 ... November(1) 97 ...
August 97 ... Previous Issues

What is the Scholars' Centre?
Opening Hours ... Introduction ... Services ... Keycard
System ... Staff ... Collections ... Databases available at
the SC
Guide to Electronic Resources for Research ... Using the
Internet for Research

The Electronic Text Service
NEW! UWA Staff and Students can now access a range
of classic texts and search their contents by keyword

Document Done

FIGURE 3.5. Web Page Viewed with Lynx (University of Western Australia Library)

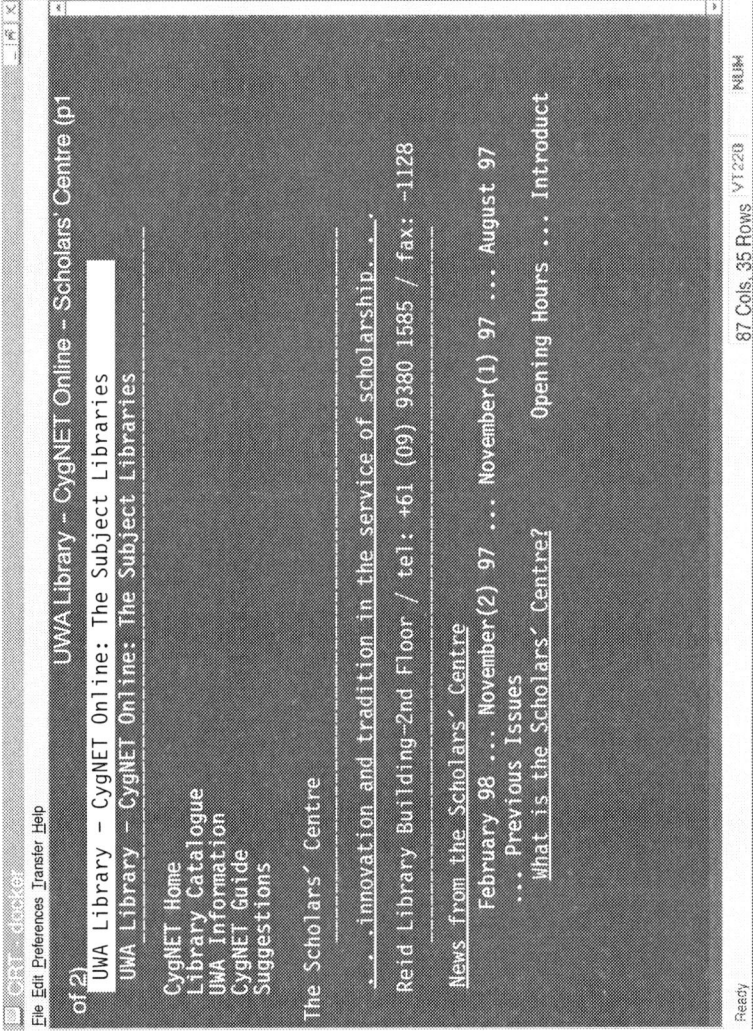

Web-browsing software is not only used with texts made available over the Internet. Some texts on CD-ROM are now designed for use with Web-browsing software such as Netscape, although they do not necessarily include a copy of it. Recent CD-ROM editions of the *Encyclopaedia Britannica*, for instance, require a Web browser to use them. This approach was presumably used partly because the *Britannica* was already available as a Web site and partly because of the ease with which images and other non-textual media can be accommodated in a Web browser.

Netscape, which is by far the most popular browser, exists in various versions, but the earlier ones do not necessarily have the full functionality needed to display files that use all the features of the latest version of HTML. Web pages with frames, forms, or embedded programs such as Java applets, for example, may not be displayed correctly—or at all—by the earliest versions of Netscape. Similar problems arise with other browsers that have gone through multiple versions. Browsers also vary in their support for particular features of HTML. This was especially the case a few years ago when Netscape and Internet Explorer were developing their own proprietary tags for extending HTML, and Web pages would be written for one browser rather than another. The introduction of HTML 4.0 is likely to mean the beginning of the end of this particular problem.

Browsers are only one side of Web software. The other side consists of the programs used for creating Web pages. There are hundreds, perhaps even thousands, of different programs of this kind, ranging from free and uncopyrighted software to very expensive commercial packages. Many of these can be used to create an electronic text that is available on the Web. They can be considered in three groups: stand-alone HTML editors, stand-alone converters, and extensions to other programs.

Stand-alone HTML editors are programs that enable the creation of HTML documents through a graphical interface, using such tools as pop-up menus and graphical tags to simplify the process of encoding the text. In some cases, they will also check that a tag is being entered in the correct context. There are various widely used commercial HTML editors of this kind, such as Adobe PageMill, Microsoft FrontPage, and SoftQuad's HoTMetaL Pro, as well as a range of free and shareware programs. They often include useful additional features such as "drag and drop" for text from other files, document templates, and tools for creating forms. Another helpful feature, offered by packages such as PageMill, is a built-in FTP program, which enables completed files to be transferred directly to the Web server on which they will reside.

Stand-alone converters exist solely to convert a file into HTML from a different format. These include RTF (Rich Text Format), Microsoft Word, PageMaker, and so on. There is even a converter from Microsoft Excel to HTML. These are usually free or shareware programs, and the HTML they produce is likely to be fairly rudimentary. They can be very useful, however, as a first pass at producing an HTML document from an existing file, backed up by subsequent reviewing and editing.

Extensions to other programs include a variety of add-ins, scripts, filters, and macros written to enable existing word-processing programs to create HTML documents. With popular word processors such as Claris Works, WordPerfect, and Microsoft Word, this means being able to save a document in HTML format. But it also may mean that access to the HTML encoding itself is not possible. The HTML tags cannot be edited directly, and the user is entirely dependent on the program itself. This may suit novice HTML authors, and those who have no interest in learning the HTML tags, but it is likely to be less acceptable to anyone who wants some control over the encoding of the text. HTML extensions to UNIX text editors, in contrast, usually involve the ability to use commands to insert tags, either from the keyboard or from a menu, without having to type in the tags themselves.

One of the essential criteria for evaluating the many different types of HTML-authoring software is the version of HTML that they support. Most programs take some time to catch up with changes to the HTML standard. Another important consideration is the extent to which they support the more sophisticated features of the markup language. Can they handle frames, for instance, and how is this done? Tables are another potential problem area: How easy is it to create tables and import data into them? Can media objects—images, sounds, movies, Java applets—be integrated into the HTML files? Above all, how does the software deal with Web "forms"?

These forms consist of a box or a series of boxes on a Web page into which a user can type information for communicating with the Web server which contains that page. One of their most important functions is to act as a conduit for entering queries or search terms. Such forms allow users to search through a text or corpus of texts for specific words or phrases, one of the most important ways in which an electronic text is different from a printed text. They are essential if the unique value of an electronic text is to be exploited satisfactorily over the Web. This function was not available in the earliest Web browsers, nor in Gopher software, but was made possible by the development of the <FORM> tag in later versions of HTML.

The input from these forms is processed by programs known as CGI (Common Gateway Interface) scripts. CGI scripts can be written in a variety of programming languages, including C and C++, Pascal, Fortran, and Visual Basic. However, the most popular method is to use the Perl language. There are various reasons for this: Perl is freely available and comparatively easy to use, and it can perform a variety of tasks. It was developed for UNIX machines but will also run on Macintosh and Windows NT Web servers. The scripts themselves exist as plain text files. There are now many publicly available collections of such scripts, designed for a wide range of different purposes. The power of CGI scripts in languages such as Perl is that they are understood and carried out by Web client-server software. Together with forms, they are essential for the effective use of Web files in HTML and other formats.

Image File Software

Web-browsing software such as Netscape may also be used to view image files directly, as long as they are stored in the GIF or JPEG formats. Many of the images provided over the Web employ this approach, including those emanating from various electronic text projects. These Web images may well be derived from other formats that are stored off-line. The sample images from the *Electronic Beowulf Project*, for instance, are presented in GIF format over the Web, while the original scans are stored in TIFF format.

The University of Virginia's Electronic Text Center provides JPEG images for use on the Web, but the original scanning is done as TIFF files, which are stored off-line on writable CD-ROMs. The *William Blake Archive* also converts its images to JPEG format for Web viewing. The *Making of America* project uses TIFF files that are converted to GIF format for Web viewing as the user requests the file (see Figure 3.6).

If an image of higher resolution is to be distributed over the Web, however, more specialized viewing software is required. The PDF format is widely used by commercial publishers for their electronic journals, but is less likely to be preferred for electronic text projects. Some of the nineteenth-century journals in the *Australian Coopera- tive Digitisation Project* contain PDF image files (see Figure 3.7). To view a PDF file requires Adobe Acrobat viewing software, known as the Acrobat Reader. Although this is widely available free of charge, it must be installed separately from, and in addition to, the Web browser. If the TIFF image format is used to provide even higher resolution, specialized image manipulation software is nec- essary before the file can be viewed.

For scholars who are interested in distributing an electronic text in image form, an extensive variety of programs is available for creating and manipulating images. The most widely used of these are probably commercial products such as Adobe Photoshop. Avail- able for all the main operating systems, Photoshop has an extensive list of features and functions, among them:

- accepting scanned images in a variety of formats;
- adjusting and correcting the color, brightness, and contrast;
- filtering the image in numerous different ways;
- transforming the image by rotating, skewing, or distorting it;
- zooming in on the image, up to 1600 percent;
- adding a digital watermark to the image; and
- tools for drawing, painting, and retouching.

As well as commercial packages such as Photoshop, however, there are many surprisingly sophisticated free or shareware pro- grams. In addition to importing images captured from scanners or digital cameras, these programs allow extensive editing of images in terms of their colors, content, size, and so on. Among the best-

FIGURE 3.6. *Making of America:* TIFF Image File Converted to GIF for the Web

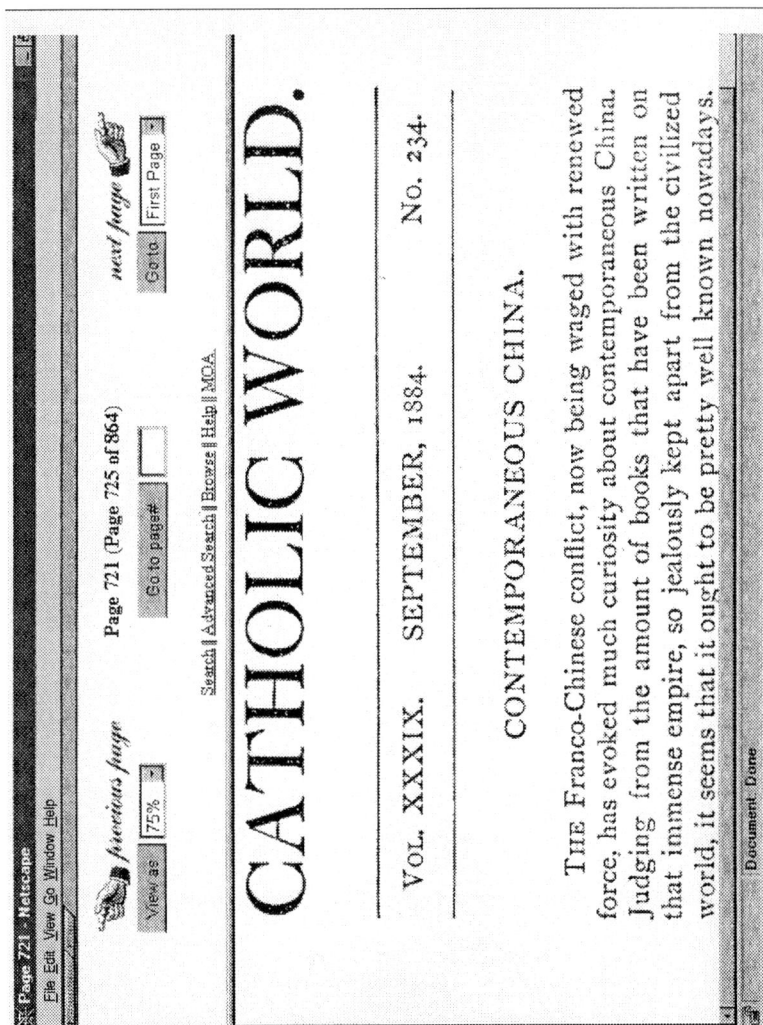

Page 721 - Netscape
File Edit View Go Window Help

previous page   View as 75%

Page 721 (Page 725 of 864)   Go to page# [   ]

next page   Go to First Page

Search | Advanced Search | Browse | Help | MOA.

CATHOLIC WORLD.

VOL. XXXIX.   SEPTEMBER, 1884.   No. 234.

CONTEMPORANEOUS CHINA.

THE Franco-Chinese conflict, now being waged with renewed force, has evoked much curiosity about contemporaneous China. Judging from the amount of books that have been written on that immense empire, so jealously kept apart from the civilized world, it seems that it ought to be pretty well known nowadays.

Document: Done

FIGURE 3.7. Australian Cooperative Digitisation Project: PDF File

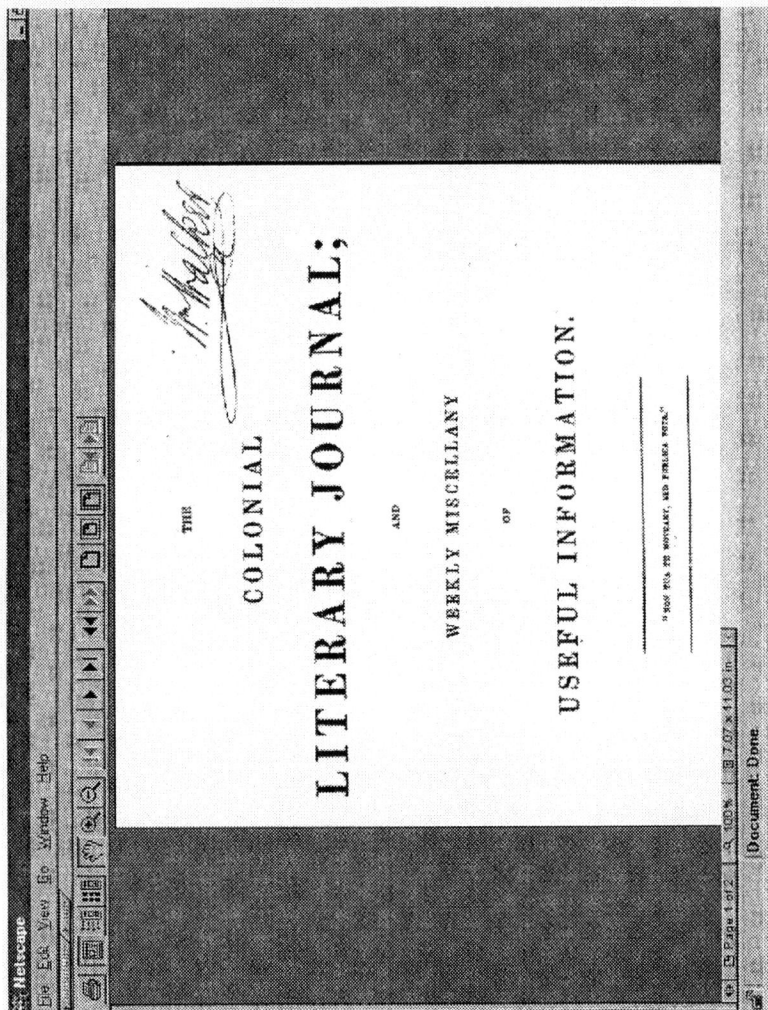

known products of this kind are XV, widely used on UNIX systems, and Paint Shop Pro, for Windows systems. A particularly good choice is ImageMagick, which is freely available and runs on Macintosh, Windows, and UNIX systems. As some of its features, it can resize, rotate, and sharpen an image, as well as altering the colors and adding special effects. Several separate images can be combined into a composite one. A small, "thumbnail" version of an image can also be created and embedded in a Web page.

Another very important property is the ability to convert image files from one graphics format to another. ImageMagick and Adobe Photoshop can import images in any one of the more popular formats, including GIF, JPEG, TIFF, and PhotoCD, and convert them to another of these formats. Various types of specialized image conversion software are also available for this purpose. For example, Graphic Converter, a Macintosh shareware program, is widely used. A more commercial product such as Equilibrium's DeBabelizer, available for Windows and Macintosh, will read and write more than 100 different types of image files.

ImageMagick includes the capacity to add annotations to images. More specialized software has also been developed for this purpose by the Institute for Advanced Technology in the Humanities at the University of Virginia. Inote, which is written in the Java programming language, enables editors to annotate images of texts and other visual objects. When these annotations are searched to identify relevant images, the image file opens with a zoomed view of the specific area of the image containing the subject of the search.

Creating an image file in PDF form will require Adobe Acrobat software. Available for both Windows and Macintosh, Acrobat can create PDF images from scanned documents and can import and convert other types of image files, including GIF and TIFF. It can also be integrated with Microsoft Word to produce PDF files from word-processor documents. As well as being viewed with the free Acrobat Reader, PDF files can be searched with the Acrobat Search utility.

Electronic Text Software

Although generic Web-browsing software may be the most convenient means of viewing and using electronic texts, it is almost certainly necessary, when creating and manipulating electronic texts, to use software designed specifically for this purpose. Such software comes in a wide variety of guises, aimed at all the different stages in the process of composing, editing, converting, storing, and publishing electronic texts. Most of it is capable of handling SGML-based documents. In fact, an extensive *SGML Buyer's Guide* (Goldfarb, Pepper, and Ensign, 1997) reveals that there are more than 150 different packages with SGML capability. Such is the complexity of the overlapping functions of these packages that the *Buyer's Guide* offers its own sophisticated methodology, known as HARP analysis, for matching the software to the specific application.

The best starting point is the many varieties of SGML editing software, which are designed specifically to create SGML-based texts. Among the best-known is SoftQuad's Author/Editor, which offers document templates with menus, spell checking, and on-line help. Other widely used editors include ArborText's Adept-Editor, Adobe's FrameMaker+SGML, and Grif's SGML Editor. The advantage of such programs is that they simplify marking up a text. Instead of having to type in all the tags by hand, you can use pull-down menus and word-processing capabilities to embed markup in a text. In fact, several popular word-processing programs offer add-ons that provide some kind of SGML capability. Corel's Word-Perfect is perhaps the best of these, but Microsoft Word also has add-ons called Near and Far Author and SGML Author. These enable SGML markup to be added to a text with the same word processor you use for "ordinary" documents.

These add-ons may have to be used with care at times, however. They have a tendency to hide the markup and to leave some of their own proprietary formatting embedded in the text file. If you want to see a text with its full, unvarnished SGML markup, and nothing else, it may be necessary to use a UNIX text editor such as Emacs with its PSGML add-on mode, which is also available for Windows and DOS. With PSGML, it is possible to automate much of the process of inserting SGML markup, while retaining full control over the process of formatting. Bob DuCharme's *SGML CD* (1998) includes a good guide to using Emacs and PSGML. The drawback is that Emacs, similar to any UNIX text editor, seems clumsy and awkward to use for anyone brought up on the popular word processors.

An important advantage of using SGML-based editing software is that it will usually validate your use of SGML for a specific text, either as markup is being entered or at the end of the data entry process. This is done by analyzing the markup in relation to the rules contained in the Document Type Definition and reporting any cases in which the markup infringes the rules. This process is known as parsing the data. Software for parsing SGML documents is widely available, sometimes as a stand-alone application, but more often as part of a larger package. The public domain parser known as nsgmls is a well-known example; it can be run by itself on UNIX, Windows, or DOS machines but is also employed by editing software such as Emacs and PSGML, among others.

One important capability relevant to the creation of electronic texts involves the conversion of so-called "legacy data"—texts which already exist as electronic files but which are not in an SGML format. To convert these requires specialized software, unless one is willing to go through them character by character, adding SGML-based markup, and stripping out proprietary formatting if necessary. An example of this kind of software is Inso's DynaTag, which will convert from various popular word-processing programs to SGML-based DTDs. It uses a batch process based on conversion templates. A more powerful and complex alternative is OmniMark, which offers a high-level programming language for manipulating SGML-based data and for converting non-SGML data to an SGML format. Other approaches to this task of conversion include using UNIX utilities such as sed and awk and using a scripting language such as Perl. These are generally more simple and freely available, but they lack any inherent understanding of the rules of SGML. To them, the output is just a string of characters, while software such as OmniMark can recognize and validate SGML-based data as well as converting them.

Once a text has been created in SGML form, or converted to it, there are various tools for adding value to the text in different ways: indexing it, searching it, delivering it to users in electronic form, and printing it out. Two packages that provide most of these features— Open Text's PAT and Inso's DynaText—are widely used in humanities computing. Both are expensive commercial products. They work by taking an SGML text (known as the "source file") and creating an index from it. The extent of this index can be customized, ranging from indexing every word to excluding a list of specific words. Preliminary style sheets are also constructed during this process, which cover the layout and appearance of the text and its table of contents. They can then be edited on-line. A style sheet to format the text for printing can also be developed.

The "books" that result from these processes can be browsed over a local area network, using a proprietary browser. But both packages also offer the capacity to develop a Web view of their texts and to serve them up onto Web browsers. In the case of Inso, this requires a complementary product called DynaWeb, which creates HTML pages from the SGML text "on the fly"—in other words, as and when the browser requests them (see Figure 3.8). The initial versions of these Web interfaces were somewhat simplistic in comparison with the proprietary browsers, but they have been greatly improved in more recent releases.

The DynaText and Open Text products also enable sophisticated searching of their texts. The elements and attributes in the SGML-based markup are used to structure and limit searches—the more sophisticated and detailed the markup, the more complex the searches can be. With DynaText, there is a choice of methods. The user may type in a search, ranging from the simple to the highly complicated, or the publisher of the text may provide ready-made searches tailored to the specific features of that text. There is also a toolkit that enables programmers to customize the appearance of the screen and the functionality of the proprietary browser, allowing even more elaborate searches. This can be seen in the customized versions of DynaText developed by Chadwyck-Healey Ltd. for its CD-ROM databases. The same company's Web services, such as *Literature Online*, are based on customized Open Text software.

It should be noted, though, that the Open Text Corporation has recently moved away from its focus on this SGML-based software, originally developed under the names PAT and Lector. Instead, Open Text is focusing on its Livelink Intranet product, an internal Web-based system for serving up documents in a variety of formats, which also offers tools for managing collaborative project work across an organization. The search engine previously used for SGML-based texts has become one component of this new product.

FIGURE 3.8. DynaWeb Publication (University of Western Australia Library)

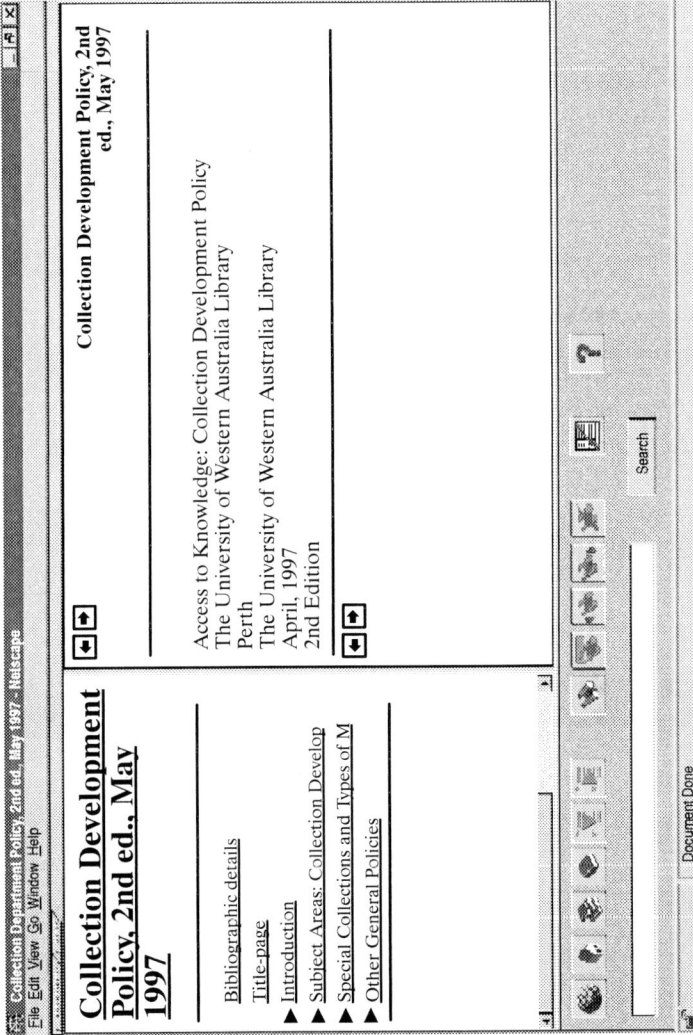

Even without access to expensive and complex products such as DynaText and Open Text, it is still possible to make use of SGML markup to search texts. With sufficient expertise, techniques such as CGI scripts and simple programming languages such as Perl can be adapted to perform searches, presented to the user as input boxes or forms on a Web page. A method for converting the results of the search from full SGML to HTML will also be needed. This is unlikely to work effectively with larger texts, however, if the source text is searched directly. To search a large text efficiently will normally require the creation of a separate index as the first step, against which searches can then be run. As is the case with Dyna-Text and Open Text, this means that any correction or alteration to the text itself will necessitate the rebuilding of the complete index.

In addition to the different kinds of software that are capable of handling texts encoded to full SGML standards, there are a rapidly growing number that will accept XML texts. Despite the recent emergence of XML, ten true XML packages were already available in late 1997 (Light, 1997:301-303), and various others were under development. Their functions include parsing XML files, translating them into other formats, searching them, and even viewing them. Many SGML products are also being adapted to support XML as well. Internet Explorer 4.0 supports an XML application—Channel Definition Formats—while several SGML editing and software development packages have announced XML capability.

Most of the SGML-based software in wide use at the moment is designed for all kinds of texts, not specifically for texts in the humanities. There are some areas of textual scholarship in which this generic software proves to be inadequate for the specialized requirements of scholars. Perhaps the most obvious of these inadequacies is in dealing with varying versions of same text, particularly as found in medieval manuscripts. Peter Robinson, of Oxford and De Montfort universities, has developed software, known as Collate, to address this area of need. Currently available only for Macintosh computers, Collate can deal with up to 100 different transcriptions of a text. The user can set the base text, and the program will then collate the other versions of the text against it. This is an interactive process, and the user can intervene to alter the collation or to regularize a word or phrase at any point. Collations based on different base texts can be compared. The collation can be output in several different styles, among which is a TEI version with hypertext capabilities. The effectiveness of Collate is particularly well demonstrated in the edition of Chaucer's *Prologue* to *The Wife of Bath's Tale*, edited by Peter Robinson and published by Cambridge University Press (Chaucer, 1996).

There are other kinds of software designed specifically for the creation and use of electronic texts, which are not based around SGML. For the most part, these predate both the development of the TEI and the emergence of software such as DynaText and Open Text. Among the more widely used examples is TACT (Text Analysis Computing Tools), developed at the University of Toronto in the 1980s, which consists of a group of fifteen programs for MS-DOS only. It is intended for individual texts, or smallish groups of texts, in languages that use the roman alphabet—though it can also handle classical Greek. Texts must be in an ASCII file, and can be marked up manually or by utility programs that indicate the dictionary form (lemma) and part of speech of each word. The text is then converted into a database, which can be searched for occurrences of a word, a combination of words, or a word pattern, and the results are displayed as a concordance, a list, or a table. Statistical analysis of the frequencies of letters, words, or phrases can also be carried out. More recently, TACT has been adapted so that it can be run as a CGI script through the World Wide Web. This enables a TACT database to be searched remotely using Web forms. TACTweb, as this software is known, is freely available for publishers of electronic texts. In this situation, users of TACT texts only need Web-browser software rather than the earlier TACT client software. (See Figure 3.9 for a sample TACTweb search.)

Another non- or pre-SGML program for textual analysis is Micro-OCP. It produces a range of analytical products—concordances, indexes, and word lists—from texts in a variety of languages and alphabets, including Arabic and Chinese. Micro-OCP is published by Oxford University Press and is only available for IBM-compatible PCs running MS-DOS. It can handle plain ASCII texts only, but can be instructed to recognize a variety of character sets. If texts contain SGML tags, these can be either ignored or listed in the output. Micro-OCP produces three kinds of lists:

FIGURE 3.9. TACTweb: Sample Search

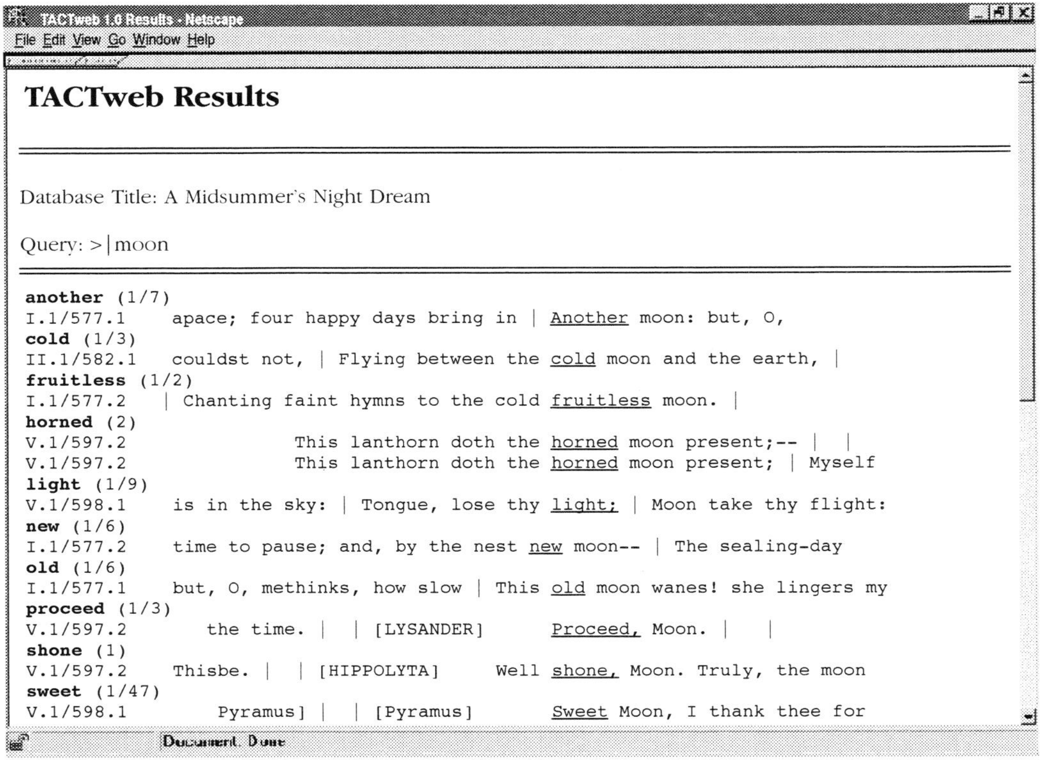
TACTweb 1.0 Results - Netscape
File Edit View Go Window Help
TACTweb Results
Database Title: A Midsummer's Night Dream
Query: >|moon
another (1/7)
I.1/577.1 apace; four happy days bring in | Another moon: but, O,
cold (1/3)
II.1/582.1 couldst not, | Flying between the cold moon and the earth, |
fruitless (1/2)
I.1/577.2 | Chanting faint hymns to the cold fruitless moon. |
horned (2)
V.1/597.2 This lanthorn doth the horned moon present;-- | |
V.1/597.2 This lanthorn doth the horned moon present; | Myself
light (1/9)
V.1/598.1 is in the sky: | Tongue, lose thy light; | Moon take thy flight:
new (1/6)
I.1/577.2 time to pause; and, by the nest new moon-- | The sealing-day
old (1/6)
I.1/577.1 but, O, methinks, how slow | This old moon wanes! she lingers my
proceed (1/3)
V.1/597.2 the time. | | [LYSANDER] Proceed, Moon. | |
shone (1)
V.1/597.2 Thisbe. | | [HIPPOLYTA] Well shone, Moon. Truly, the moon
sweet (1/47)
V.1/598.1 Pyramus] | | [Pyramus] Sweet Moon, I thank thee for
Document: Done

- A concordance that lists each word that occurs in the text, together with the number of occurrences, the location reference of each occurrence, and its textual context.
- An index that lists each word occurring in the text, with its frequency and location references.
- A list of words, giving frequencies for each.

A somewhat different set of functions is offered by CASE (Computer-Assisted Scholarly Editing) (Shillingsburg, 1996:141-144). Its main purpose is to collate lengthy prose texts. Transcriptions of up to eight different versions of a text can be entered and collated automatically. The result is a copy text accompanied by files of variants, which can be examined and used to produce an emended text for a critical edition. The variants can be conflated into a single file, from which an editorial apparatus can be derived. To convert these materials for typesetting or electronic publication requires the addition of appropriate encoding, such as SGML-based markup or TeX codes. This can be either added at the end of the CASE process (manually or semiautomatically) or produced by inputting short mnemonic codes along with the text, with a conversion program applied afterward. CASE will handle the collation of manuscripts as well as printed works. It runs on various platforms, including MS-DOS and Macintosh. The Macintosh version is designed as a series of HyperCard stacks.

TUSTEP is a rather similar package that has been developed over more than twenty years and was used by Hans Walter Gabler in preparing his 1984 edition of James Joyce's *Ulysses*. It can now carry out a wide range of functions in textual analysis:

- Importing text from OCR and word-processing programs
- Collating and listing variations in different versions of a text
- Batch updating of texts
- Indexing and sorting words or lexical units within a text
- Preparing output for printing and typographic composition

A recent addition has been better support for texts encoded with SGML or XML. Though TUSTEP itself is bilingual in English and German, most of the documentation and training is only available in German.

Packages such as TUSTEP are essentially university-developed products that tend to emphasize analytical functions such as collation. Commercial packages for publishing electronic texts are rather different in their emphasis. One important non-SGML product is the Folio Views and Folio Builder family of software, which is widely used in the corporate and government sectors. It works by converting documents from various formats—including SGML, HTML, Word, Word-Perfect, and RTF—into a proprietary "infobase." Searches and queries are supported, and the ability to embed hyperlinks within the text is included. Using a Folio-based text requires either the proprietary viewing software (Folio Views) or an ancillary product (Folio SiteDirector) that converts the text for viewing on the World Wide Web with a generic Web browser. The crucial difference between Folio Views and a product such as DynaText is that Folio does not use SGML as its internal standard, preferring its own proprietary format instead. The only major example of scholarly electronic texts that use Folio Views is Intelex's *Past Masters* series of philosophical works. The original diskette version of this series came with Folio Views, as did the later CD-ROM version. The same texts are now offered over the Web, using the conversion tool to allow access by Web browsers (see Figure 3.10).

Another product in this area is the IBM Digital Library software, which provides access to a collection of "multimedia objects" through various linked servers. A library server holds a database with metadata, or catalog information about the collection, while one or more object servers hold the digital objects themselves. Access for users is through proprietary client software or through standard Web browsers. A variety of searches can be run against the data, including natural language searching of the full text. An important feature is the level of security that can be applied to the collection, including watermarking for images and the ability to control authentication, encryption, and payments systems. Nevertheless, the IBM Digital Library does not recognize SGML-based markup as such and lacks the ability to link searching to the tag structure of a text.

Also worth mentioning is a program that was the ancestor of many of the features embodied in the World Wide Web. Apple's Hyper-Card first appeared in 1987 and probably reached its peak in the

FIGURE 3.10. FolioViews: Web Version (Machiavelli, *Il Principe*)

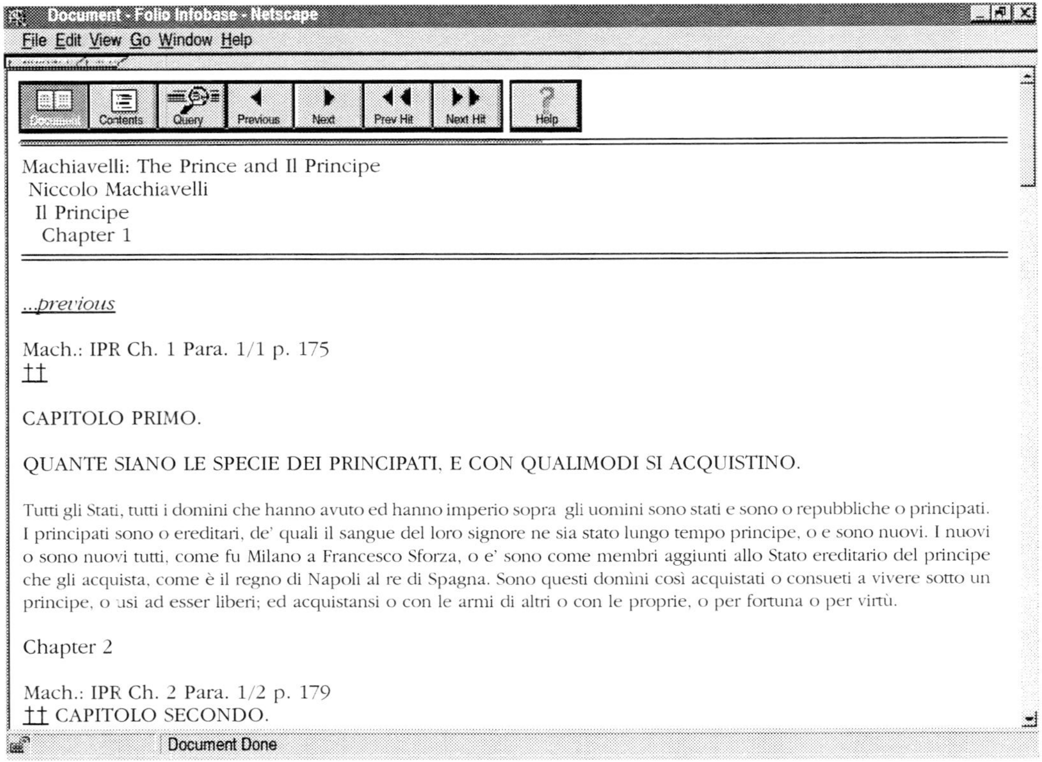
Document - Folio Infobase - Netscape
File Edit View Go Window Help
Document
Contents
Query
Previous
Next
Prev Hit
Next Hit
Help
Machiavelli: The Prince and Il Principe
Niccolo Machiavelli
Il Principe
Chapter 1
...previous
Mach.: IPR Ch. 1 Para. 1/1 p. 175
††
CAPITOLO PRIMO.
QUANTE SIANO LE SPECIE DEI PRINCIPATI, E CON QUALIMODI SI ACQUISTINO.
Tutti gli Stati, tutti i domini che hanno avuto ed hanno imperio sopra gli uomini sono stati e sono o repubbliche o principati. I principati sono o ereditari, de' quali il sangue del loro signore ne sia stato lungo tempo principe, o e sono nuovi. I nuovi o sono nuovi tutti, come fu Milano a Francesco Sforza, o e' sono come membri aggiunti allo Stato ereditario del principe che gli acquista, come è il regno di Napoli al re di Spagna. Sono questi domini così acquistati o consueti a vivere sotto un principe, o usi ad esser liberi; ed acquistansi o con le armi di altri o con le proprie, o per fortuna o per virtù.
Chapter 2
Mach.: IPR Ch. 2 Para. 1/2 p. 179
†† CAPITOLO SECONDO.
Document Done

early 1990s, when it was being distributed with all new Macintosh computers. HyperCard is built around the metaphor of a stack of cards that can be interlinked in numerous ways and browsed in a variety of directions. It also contains tools for importing and manipulating images and sound and video files, as well as for drawing and painting. External processes and programs can be controlled from within the HyperCard structure. Other features include a programming language called HyperTalk and the capacity to create indexes and carry out searches. HyperCard stacks can now be made available on the World Wide Web using software called LiveCard.

HyperCard was used for a range of experimental electronic texts in the late 1980s and early 1990s, including some French emblem books of the sixteenth century (Graham, 1991) and Fielding's *Joseph Andrews* (Delaney and Gilbert, 1991). Its most successful use was by the *Perseus Project*, which issued two HyperCard versions of its Greek digital library on CD-ROM in 1992 and 1996 (Crane, 1998). Although HyperCard now appears to have been largely abandoned for this kind of scholarship, it has played an important role in developing ideas and approaches for multimedia editions of literary texts.

Delivering the Software

When an electronic text is delivered to its user, the publisher may choose either to distribute it without any software, or to bundle the software with the text. Texts published commercially on CD-ROM usually follow the latter approach. Chadwyck-Healey Ltd. includes a version of the DynaText browser software with its CD-ROM publications. Cambridge University Press does the same for its CD-ROM texts. Intelex publishes its *Past Masters* CD-ROM of philosophical texts with the Folio Views software. Some texts, such as the recent CD-ROM editions of the *Encyclopaedia Britannica*, include Web-browsing software instead. The *Oxford English Dictionary* on CD-ROM, in contrast, comes with its own custom-made software, designed by AND Software of Rotterdam. So do several other publishers' products: the *Cetedoc Library of Latin Texts*, published by Brepols, includes software by Dataware and Orda-B, while the *Hartlib Papers* CD-ROM set includes software known as TOPIC, produced by the California-based company Verity Inc.

When these same texts are supplied on magnetic tape for local networking, however, they generally come without any software at all. Both Chadwyck-Healey and Intelex only supply the SGML source files. In the case of the former, they are accompanied by their own DTD files. Intelex, however, uses the TEI Lite DTD and does not distribute it with the source files. The magnetic tape version of the *Oxford English Dictionary* is quite unusual; the source files, which are in an SGML-like format, come with index files for the Open Text software. There is no DTD, although one is available privately on the World Wide Web for institutions that want to network the OED using different software.

The whole aim of SGML, of course, is to enable texts to be delivered independently of any specific software. Hence, the commercial versions of texts, with their bundled software or their Web-based searches, might be considered restrictive. They allow no access to, or manipulation of, the source files, and they provide sophisticated but "canned" searches of the texts. This is very helpful and worthwhile for their users, particularly if local expertise and resources are limited. But for the scholar who wants to customize the way a text is presented for browsing and the way in which it can be searched, these commercial products can have their limits. The alternative is to acquire or create SGML-based source files and to process and deliver them locally using one's own choice of software. This enables a far greater degree of customization. It does, however, require considerable local expertise in the design and use of appropriate transmission media and software for delivering electronic texts.

FURTHER READING

An exhaustive guide to software and companies offering SGML-related services is the *SGML Buyer's Guide* compiled by Charles F. Goldfarb (inventor of SGML), together with Steve Pepper and Chet Ensign (Goldfarb, Pepper, and Ensign, 1997). It covers more than 150 different kinds of software and also provides a formal methodology, known as HARP analysis, for analyzing work flows and software requirements for electronic publishing and document management. Travis and Waldt (1995) cover much of the same ground in a concise, narrative form.

Bob DuCharme's book (1998) and CD-ROM, *SGML CD*, contains a full range of public domain SGML software, with detailed instructions on its use.

The definitive on-line guide to SGML-based software is Steve Pepper's *Whirlwind Guide to SGML Tools.*

More specialized treatment of specific software can be found in Burrows (1996), which describes the use of DynaWeb for delivering large-scale electronic texts in the humanities, and Robinson (1994:20-23), which gives an account of Collate and its application to the CD-ROM edition of the *Canterbury Tales.*

There are numerous books on CGI scripts and Web servers and quite a range on the Perl programming language in particular. See, for example, Palmer, Schneider, and Chenette (1996); Quigley (1995); and Wall (1996).

On TACT, see Lancashire and others (1996).

There are many guides to specific HTML authoring tools. These include McClelland and San Filippo (1997) on PageMill and Matthews and Poulsen (1998) on FrontPage.

A useful survey of trends in the Web-browser market is given by Karpinski (1997).

WEB SITES

Adobe products: http://www.adobe.com/

Center for Computer Analysis of Texts, University of Pennsylvania:
   gopher://ccat.sas.upenn.edu:3333/00/re.ccat

Collate: http://slate.blue.dmu.ac.uk/projects/Collate/

Dartmouth Dante:
   http://www.nyu.edu/library/bobst/research/etc/dante.htm

DynaText & DynaTag (Inso Corporation): http://www.inso.com/

Electronic Beowulf Project:
   http://www.uky.edu/~kiernan/BL/kportico.html

Folio Views: http://www.nlx.com/pstm/pstmsoft.htm

HyperCard Resource Page: http://www.glasscat.com/hypercard/

IBM Digital Library: http://www.software.ibm.com/is/dig-lib/

ImageMagick:
  http://www.wizards.dupont.com/cristy/ImageMagick.html

Intelex: http://www.nlx.com/pstm/index.html

Literature Online (Chadwyck-Healey): http://lion.chadwyck.com/

MacCASE: http://www.adfa.oz.au/ASEC/

Micro-OCP: http://www1.oup.co.uk/E-P/Humanities/Micro-OCP/

Microsoft FrontPage:
  http://www.microsoft.com/frontpage/productinfo/overview.htm

OED DTD: http://weber.u.washington.edu/~dillon/engl569/oed/

Online Book Initiative: http://world.std.com/references/obi.html

Open Text: http://www.opentext.com/

Panorama viewer:
  http://www.softquad.co.uk/products/panorama/panvfeat.html

PLD Web (Chadwyck-Healey): http://pld.chadwyck.co.uk/

Project Gutenberg: http://promo.net/pg/

SoftQuad: http://www.softquad.co.uk/products/pc-hmp4.html

Steve Pepper, *The Whirlwind Guide to SGML Tools and Vendors:*
  http://www.falch.no/people/pepper/sgmltool/

TACT: http://www.chass.utoronto.ca:8080/cch/tact.html

TACTweb: http://tactweb.mcmaster.ca/tactweb/doc/tact.htm

TUSTEP: http://www.uni-tuebingen.de/zdv/zrlinfo/tdv_eng.html

University of Virginia Electronic Text Center:
  http://etext.lib.virginia.edu/

Virginia Tech Gopher: gopher://gopher.vt.edu:10010/10/33

William Blake Archive:
  http://jefferson.village.virginia.edu/blake/main.html

Wiretap Online Library: http://www.area.com/

Chapter 4

Organizing Access to Electronic Texts

Electronic texts, similar to printed ones, require a publishing infrastructure of some kind to distribute them and make them available to the scholarly community. There are various ways of doing this, although most of them are still in their infancy. It is not yet clear which of them will turn out to be valid in the long term.

In this chapter, we look at the different avenues that are available for organizing access to electronic texts. We begin with individual scholars publishing and distributing their own work and go on to examine various institutional approaches: electronic text centers, centers for humanities computing, so-called "digital libraries," and national and international services. The respective roles of academic and commercial organizations are also discussed.

THE INDIVIDUAL SCHOLAR AS PUBLISHER

The simplest approach to publishing an electronic text is for individual scholars to do it themselves. This can be done on diskette or CD-ROM, neither of which is particularly expensive if the basic technological infrastructure is already available. For a diskette, the minimum requirement is little more than a suitable personal computer, with appropriate software, and a supply of diskettes. The texts can then be posted directly to interested scholars around the world, handed out at meetings and conferences, or distributed through more formal publishing channels.

Publication on CD-ROM requires more resources, particularly the software and mastering equipment necessary to write to a CD-ROM. The disks themselves are comparatively inexpensive, and a publication can be duplicated for relatively little cost. Commercial bureaus will provide mastering and duplicating services. CD-ROM publications must then be distributed, and the avenues for doing this are the same as those for diskettes.

A crucial question that arises with both diskettes and CD-ROMs is which operating system to provide for. Even when a plain "ASCII" file is used for the text, the diskette itself must be formatted for either Macintosh or IBM-compatible machines. The more recent computers can usually handle diskettes of either type, but most older ones will only recognize one of these types. Publication on CD-ROM makes it easier to allow for both types of machine, but a different problem may arise: only the more recent machines will have a built-in CD-ROM drive, and many people are unlikely to have easy access to one. These kinds of considerations ought to make a scholar think twice before distributing electronic texts on diskette or CD-ROM— let alone the much more complicated factors involved in deciding whether to include any software with the text and whether to design for a particular type of program.

Concerns such as these mean that the most likely method of publication for the individual scholar is over the Internet and, usually, through the World Wide Web. The relative simplicity of the Web as a publishing medium is reflected in its enormous popularity, with tens of thousands of people around the world publishing Web pages. The Web has been called the greatest vanity press ever invented, and most of the personal Web pages exhibit some degree of self-indulgence. All the same, the Web is an excellent medium for publishing serious, scholarly work. Its additional advantage is that it is also a powerful channel for disseminating the text, without the need to distribute a physical object.

The prerequisites for publishing on the Web are a connection to the Internet, a computer running Web-server software, an IP (Internet Protocol) address for the server, and access to the appropriate directories and files on the server. Most, if not all, universities and colleges have this kind of setup, and in many cases, schools and departments have their own servers. Faculty and students should get to know their nearest Web expert and find out which Web publishing services are offered. Outside academia, such facilities are less likely to be readily available, but now numerous commercial firms do provide Internet access and will usually allow Web publishing for a fee. The availability of such services is likely to grow rapidly in the future.

It also helps to know at least the basic principles of Web page authorship and design, but even this is not absolutely essential. Other people with these skills can easily be employed if necessary, either as freelance individuals or through commercial bureaus. Some may even do the work free of charge. There are numerous semischolarly sites on the Web where the labor and expertise have been contributed by volunteers. Alternatively, the use of "Web-authoring" software such as Adobe PageMill or Microsoft FrontPage can simplify the Web-publishing process considerably, albeit at the expense of some loss of control over the details of the appearance and formatting of the Web pages.

An interesting example of an individual scholar publishing electronic texts over the World Wide Web is provided by James O'Donnell, who is a Professor of Classical Studies at the University of Pennsylvania. He has made a variety of texts available through his personal Web site, as part of a wide-ranging collection of scholarly materials (see Figure 4.1). The texts, which are all in HTML, reflect his specialized field: postclassical Latin literature. They include St. Augustine's *Confessions* (in Latin) and *On Christian Doctrine* (in English), the *Consolation of Philosophy* of Boethius (in both Latin and English), and the *Apology* of Apuleius (also in Latin and English).

These are accompanied by an eclectic "private library" containing texts and snippets of interest to him, ranging from Plato and Aristotle to poetry by Shakespeare and Rilke, as well as Joyce's *Ulysses*. Some of these are his own electronic versions, while others are pointers to sites elsewhere on the Web. All this is rounded out with copies of his own publications, including a "post-print" of his 1979 book on Cassiodorus and his 1985 book on Augustine, articles on Augustine and on postclassical Latin literature, book reviews, and teaching materials.

Sites such as O'Donnell's provide good insight into the work and interests of an individual scholar who is prepared to develop a personal corpus of this type. There are certainly advantages in publishing and distributing electronic texts directly in this manner. If the technological infrastructure is readily available, there is comparatively little cost involved. There is no need to justify or subsidize commercial publication of what are, after all, fairly noncommercial products. There is no need to wait for the often lengthy processes of

FIGURE 4.1. James J. O'Donnell (Home Page)

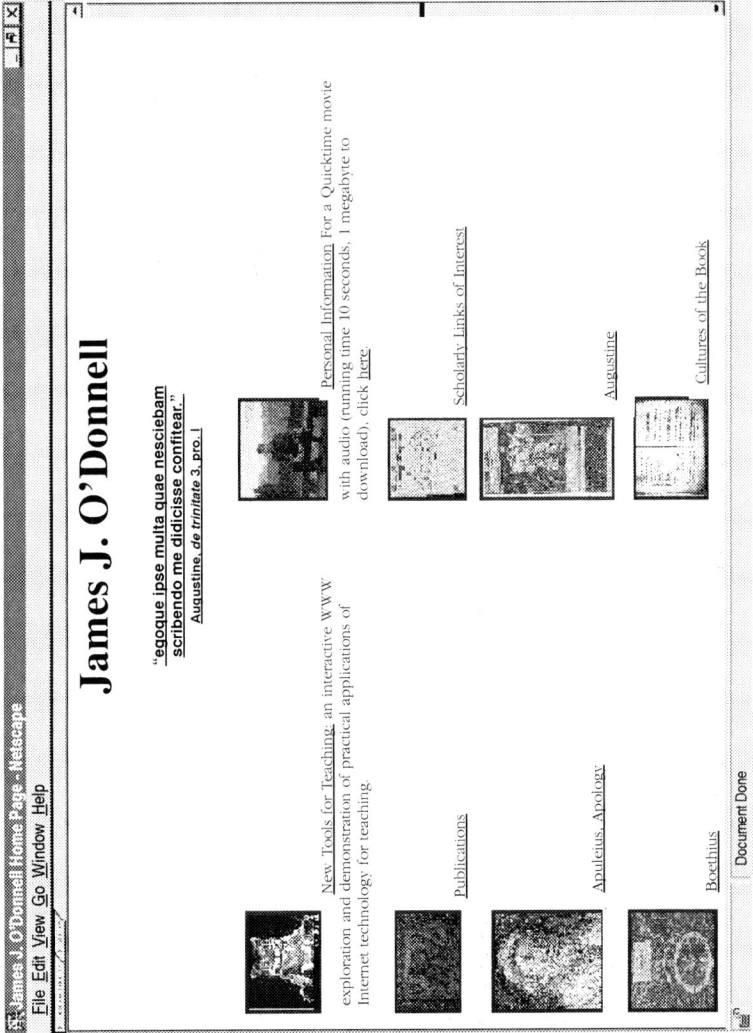

print publication. Personal Web sites, in particular, have a high degree of immediacy and accessibility.

The World Wide Web also offers the possibility of reaching a new audience. In contrast to the rarified, specialized, and narrow world of traditional scholarly publishing, the Web is perceived as a new and exciting medium. Its reach is global, and it is accessible to everyone with the basic equipment. Contact between the user of a Web site and its authors can be direct and immediate. The creators of many scholarly sites have remarked on the extent to which their work is being used by people outside the world of universities and colleges, particularly by school students and the interested amateur. As Gregory Crane observes, in connection with the *Perseus Project*, "we can see by the patterns of use and the mail we receive the stirrings of a vast audience, hungry for ideas and for that practice of thought to which we, professional academics, have been privileged to dedicate our lives" (Crane, 1998).

Although publishing by individual scholars can tap into this audience, there are several important limitations that may reduce the effectiveness of this personal approach. They are only partly connected with the need to find suitable hardware and software and to identify people with the appropriate expertise. More important, if such individual publishing is done on the basis of inadequate resources or knowledge, there is a real danger that the resulting product will be amateurish and unsatisfactory. A poorly designed Web site, for example, may seriously undermine the value and usefulness of the texts presented through it and will reduce the credibility of both the author and the material. A badly structured or over simplified version of a complex text may be seen as misleading and unhelpful. Insufficient time is also likely to be a problem for the individual scholar, who is often trying to fit the development of an electronic text into an already busy schedule of teaching and research.

A second major limitation is that self-published electronic texts have, more often than not, bypassed the usual system of peer review and "quality control." They are the electronic equivalent of the self-published book and the unrefereed journal, with no independent assessment of their scholarly quality. There are various means of addressing this question. One is to accompany the text with suitably detailed documentation of the procedures followed in its creation. James O'Donnell's Latin text of Augustine's *Confessions*, for example, carries a disclaimer acknowledging the uncertainty of its provenance. Other Web publishers are not always so scrupulous.

Another approach to quality control is to subject the text to peer review before its publication and to record the results with the text. A third approach, suitable for Web-published texts, is to provide a means for other scholars to record their comments after reading and using the text and to include these in the materials available on the Web site. Alternatively, the quality of the text may be left for reviewers to determine, as with published books, but there is little evidence yet of Web sites or CD-ROMs being reviewed in the "book review" sections of major scholarly journals. At best, electronic texts are reviewed in specialized journals dealing with computing in the humanities, and even these tend to ignore materials published on the Web.

Another significant drawback for individually published electronic texts is that they may disappear without a trace—just as self-published books tend to do without some attention to marketing and distribution. If they are to reach their intended audience, they need to be integrated into a broader framework, within which potential users are likely to find them. This is analogous to the way in which libraries and bookstores make printed material available, although the options for electronic texts are somewhat different. For CD-ROMs and diskettes, this may mean housing a copy of the text in an institutional setting, which will enable other scholars to have access to it. This may be a library, a computing center, or a more commercial avenue such as a bookstore.

A similar problem, but one with fewer precedents, arises with electronic texts published on the World Wide Web instead of on CD-ROM or diskette. It is not enough to put up a Web site and hope that interested people will find it. To communicate effectively, the site must be linked into the wider frameworks of knowledge and scholarship on the Web. This can be done by ensuring that the site is included in Web indexes, and particularly in the subject guides that relate to the specific discipline involved. Medieval texts, for example, ought to appear in the lists of links offered by sites such as the *Labyrinth Library of Medieval Studies* and *NetSerf*. Another vehicle for publicity is to announce the existence of the site on the appropriate electronic mailing lists and bulletin boards, but this is likely to be more ephemeral than a listing in a subject guide.

At the same time, suitable provision should be made for the continued availability of the text over the Web. Although texts on CD-ROM can be published in a way similar to books, it is the responsibility of the purchaser of the text, not the publisher, to make provision for its storage and for continued access to it in the future. However, Web-based texts require the publisher to ensure their availability. This will depend, to a large extent, on which computer is actually housing the public version of the text. A personal or private machine is far less preferable than a server controlled and maintained by an institutional unit with specific reponsibilities and appropriate expertise. If no local facilities of this kind are available, it may be possible to have a copy of the text housed on a remote server. The Oxford Text Archive, for instance, provides a service of this kind, acting as a clearinghouse for texts prepared by individual scholars.

If sufficient attention is paid to the electronic equivalent of distribution and marketing, an individual scholar may well be able to publish electronic texts with some success. This is likely to be considerably enhanced if a group of scholars combine to publish and distribute a text. Academic departments and scholarly associations are usually better able to find suitable resources than the individual scholar is. To begin with, the spread of expertise is likely to be greater, particularly in creating and designing Web pages and sites and in marking up texts, and, if a project can be established to bring together a group of scholars with the necessary equipment, software, and earmarked funding, so much the better. Many of the electronic texts on the Web are being developed in this way. The *Canterbury Tales Project*, for instance, is funded by the British Academy and the Leverhulme Trust and is based at the University of Oxford and the University of Sheffield.

Some of the limitations that arise with individual scholarly publishing may also apply to electronic text projects. Peer review and quality control can still be an issue, but a project committee or advisory group will usually serve as the equivalent of an editorial board and provide some guarantee of the content and validity of the texts being published. Continued availability of Web-based texts will also need to be addressed. Although most projects are likely to set up their own Web server, they usually have a limited life span. Provision must be made for the text to be kept available and maintained after the project itself has ceased work.

There are also some limitations that apply to projects, but not to individual scholarly publishing. Most projects are likely to face some differences of opinion over the standards to be used, the design and presentation of the text, and the overall strategy for the publication of the text, even when these issues have been discussed in considerable detail before the project gets under way. These differing views may slow down the project or even halt it altogether. They may also result in compromises of the "lowest common denominator" type. More innovative and original approaches will be lost.

The reliance of almost all projects on special funding also carries its dangers. Projects may be led into false expectations about how much can be achieved with limited funds, with overambitious aims being developed only to fail in their execution. Special funding also tends to reinforce an assumption that this kind of work is outside the scholarly mainstream. If additional equipment is acquired and additional staff are employed purely to carry out the project, the deduction is readily made that electronic texts can only be produced when these resources are available. For universities, there is something of an analogy here with the widely held assumption that research can only be done when research grants are available to fund it.

Whether electronic texts are prepared by project teams or by individual scholars, and whether they rely on special grants or not, an international infrastructure is essential for publishing, distributing, and preserving these texts for the academic community. Suitable institutional settings where such texts can be collected and organized for scholars to use are needed. As with the infrastructure for printed materials, this involves a mixture of commercial and noncommercial approaches. The names, roles, and locations of the institutions are rather different, though, and include such specialized bodies as electronic text centers, centers for humanities computing, and digital libraries.

INSTITUTIONAL APPROACHES TO PUBLISHING

Institutional initiatives in publishing and distributing electronic texts, especially in the wake of the enormous growth in this area since the early 1990s, have been somewhat varied. In some cases, existing structures were adapted or expanded to incorporate new responsibilities for networked distribution of texts. Some humanities computing centers have followed this path, as discussed next in this section. More usually, though, a new facility has been established, under the generic title "electronic text center." An extensive survey of twenty-two such centers in North America, published in 1996 (Smith and Tibbo, 1996), provides some useful information about their characteristics.

Electronic Text Centers

The electronic text center model has been very successful. Having the organizational and financial support of one of the larger units in a university—the library—has been beneficial. Being able to draw on the library's experience in selecting and organizing materials for access has also been an important asset. The library is more likely to take a broad and long-term view and to build a service that is widely accessible, relevant to a broad range of users, constructed in accordance with accepted standards, and preserved for access in the future.

The typical electronic text center is located in the main humanities library. It is usually a distinct, publicly accessible area, with facilities for creating electronic texts and for accessing both networked and nonnetworked texts. It has its own specialized staff who may have a mixture of expertise, covering librarianship, computing, and research in the humanities. Graduate students are often employed as well. Training in the creation and use of electronic texts is provided for faculty and students. One of the earliest and most influential examples of this kind of electronic text center has been that of the University of Virginia Library, which has been in existence since 1992 under the direction of David Seaman (see Figure 4.2).

An electronic text center ought to work closely with faculty to ensure that the texts being created and distributed are relevant to the teaching and research of the university. It also has its own teaching

FIGURE 4.2. Electronic Text Center, University of Virginia Library (Home Page)

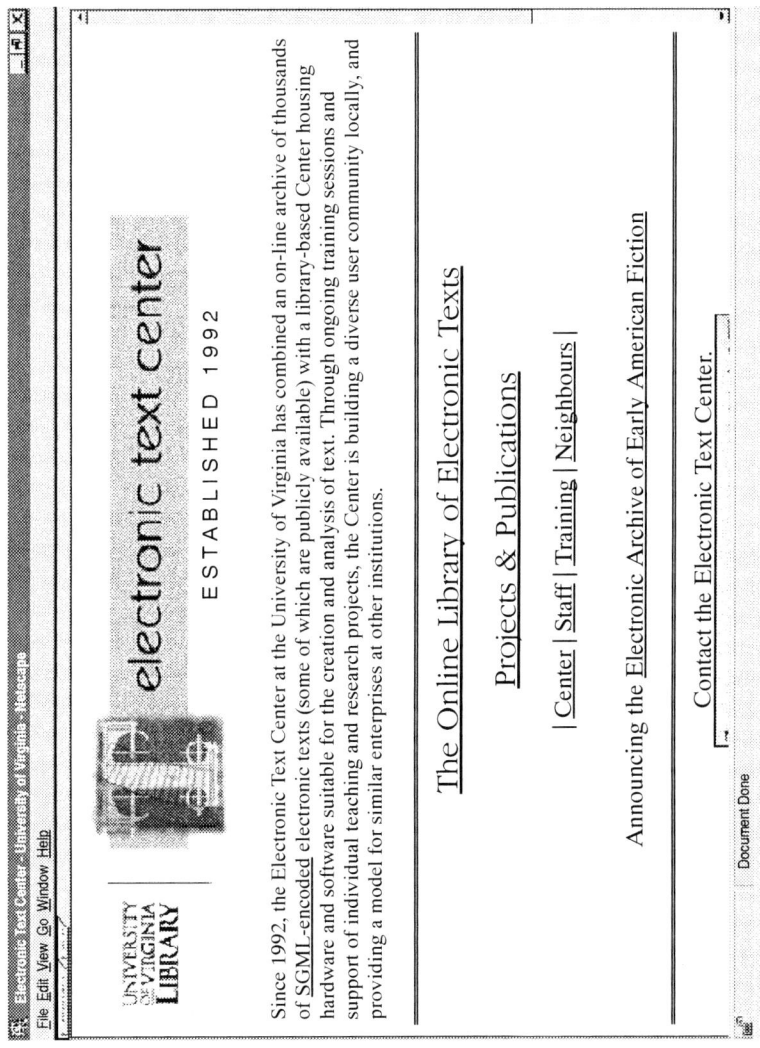

UNIVERSITY
OF VIRGINIA
LIBRARY

electronic text center

ESTABLISHED 1992

Since 1992, the Electronic Text Center at the University of Virginia has combined an on-line archive of thousands of SGML-encoded electronic texts (some of which are publicly available) with a library-based Center housing hardware and software suitable for the creation and analysis of text. Through ongoing training sessions and support of individual teaching and research projects, the Center is building a diverse user community locally, and providing a model for similar enterprises at other institutions.

The Online Library of Electronic Texts

Projects & Publications

| Center | Staff | Training | Neighbours |

Announcing the Electronic Archive of Early American Fiction

Contact the Electronic Text Center.

role. The late Anita Lowry, director of the Information Arcade at the University of Iowa, expressed this well:

> the library is the natural venue for teaching people to use electronic primary source materials for research and teaching and . . . we must prepare to take the lead in teaching ourselves and others what is possible and how to do it. (Lowry, 1997:204)

From this perspective, the electronic text center is also an electronic classroom.

Centers for Humanities Computing

An older approach, with a somewhat different purpose, is the center for humanities computing. Such centers are comparatively uncommon (Ellis [1996:528] estimates fewer than a dozen in the world), but they have existed in some universities since the 1960s or 1970s. Their original purpose was to provide a focus for early applications of computers to research in the humanities, both in offering suitable expertise and equipment and in assisting with data storage and management. One important aim was to offer a separate computing facility outside the domain of the sciences, which dominated the early use of computers in universities. A particular emphasis of the humanities computing centers was on acquiring or developing suitable software and on providing training and support for faculty. Their major areas of interest since the 1970s and 1980s have been the construction of linguistic corpora and textual concordances and the statistical analysis of texts (stylometrics).

These centers generally did not see their role as publishing and distributing electronic texts to a wider community and did not usually consider a wider clientele than the direct users in the relevant academic departments. Their role was to support these users, not to build an infrastructure for communicating with scholars in other universities. If an analogy is to be sought, it can be found in campus computing centers, which flourished in the 1970s and into the 1980s, but then, in many cases, were broken up and decentralized or merged with libraries in an attempt to create a single information service. These centers tended to lack the library's breadth of outlook and its mission of service to all, not just to a group of specialist users.

Centers of humanities computing still have an important role today, but it is changing to accommodate developments in networking. Among the services likely to be provided are Web sites and electronic mailing lists, which bring a center's expertise and knowledge to a global audience rather than a local one. Courses, seminars, and conferences remain major activities, often accompanied by publications in both printed and electronic forms. Some centers are also involved in making electronic texts available, incorporating the functions of an electronic text center. Among the best-known of the humanities computing centers are:

- the Humanities Computing Unit at Oxford University, which is linked to the Oxford Text Archive but is primarily a support service offering advice and training (see Figure 4.3);
- the Centre for Computing in the Humanities at King's College London, which has as its goal the fostering of awareness, understanding, and skill in the scholarly applications of computing;
- the Center for Computing in the Humanities at the University of Toronto, which offers training courses and holds a lecture series and is also involved in making electronic texts available;
- the Institute for Advanced Technology in the Humanities at the University of Virginia, which provides a focus for a variety of projects developing electronic texts, particularly of an experimental type;
- the Computing for the Humanities User Group at Brown University, which coordinates a variety of groups examining areas such as hypertext, markup, and scholarly technology; and
- the Norwegian Computing Centre for the Humanities (in Bergen), which focuses on training and information services (Mullings et al., 1996:xvii-xviii).

Multimedia Centers

Many universities and colleges have some kind of multimedia center, either as a central facility or within a specific school. In some cases, these centers have grown out of earlier computing or photographic units, while in others they have been established only recently. Their main focus is on the technical support and facilities needed to create electronic products involving digitized images,

FIGURE 4.3. Humanities Computing Unit, University of Oxford (Home Page)

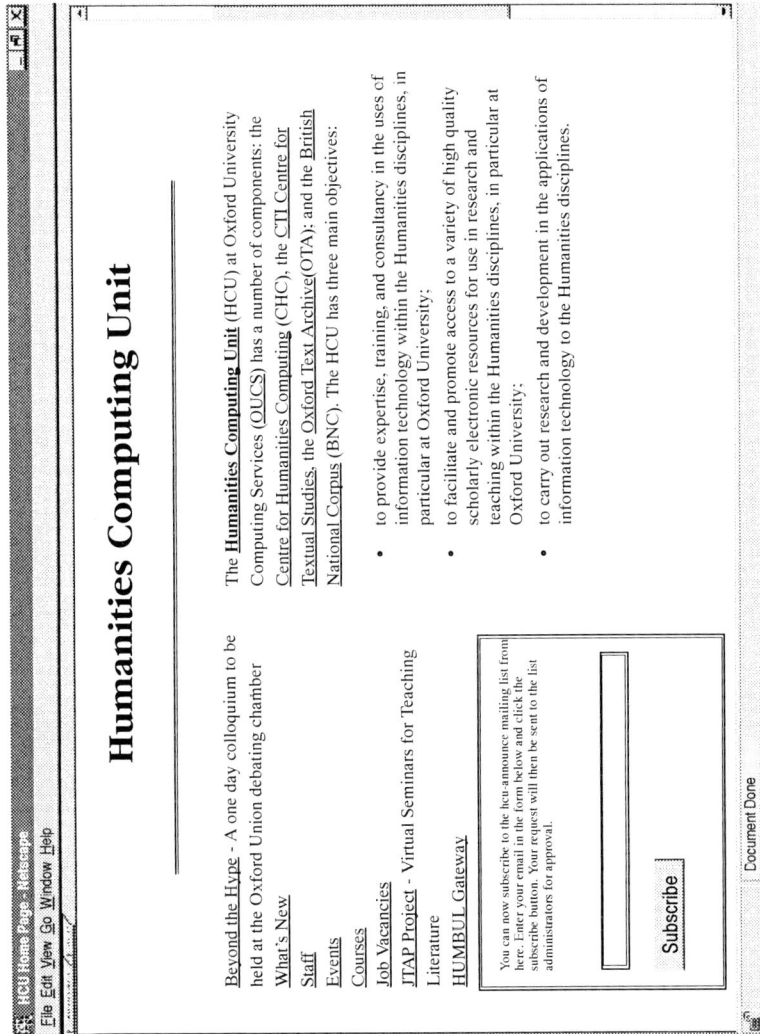

File Edit View Go Window Help

Humanities Computing Unit

Beyond the Hype - A one day colloquium to be held at the Oxford Union debating chamber

What's New

Staff

Events

Courses

Job Vacancies

JTAP Project - Virtual Seminars for Teaching

Literature

HUMBUL Gateway

You can now subscribe to the hcu announce mailing list from here. Enter your email in the form below and click the subscribe button. Your request will then be sent to the list administrators for approval.

[Subscribe]

The **Humanities Computing Unit** (HCU) at Oxford University Computing Services (QUCS) has a number of components: the Centre for Humanities Computing (CHC), the CTI Centre for Textual Studies, the Oxford Text Archive(OTA); and the British National Corpus (BNC). The HCU has three main objectives:

- to provide expertise, training, and consultancy in the uses of information technology within the Humanities disciplines, in particular at Oxford University;

- to facilitate and promote access to a variety of high quality scholarly electronic resources for use in research and teaching within the Humanities disciplines, in particular at Oxford University;

- to carry out research and development in the applications of information technology to the Humanities disciplines.

Document Done

101

sound, and video. They may provide training for users who want to make their own products, but are more likely to offer fee-based services. They may sometimes work closely with faculty in developing and presenting course materials, and their services may be incorporated to some extent into the teaching curriculum. More commonly, though, they are a service bureau offering a range of digitization services at a suitable fee.

Multimedia centers are more likely to be involved in short-term, discrete projects than in creating a coherent range of electronic resources (Ellis, 1996:528). They do produce electronic texts, as long as these are "multimedia" in some way, but they are not greatly concerned with electronic texts in the sense that we are considering them here.

Digital Libraries

The term "digital library" is now frequently used, having seemingly replaced the earlier "electronic library" or "virtual library." Unfortunately, it tends to be used loosely and generally for almost any collection of digital resources. If the term is to have any meaning, it must relate to the inclusion of the word "library." Eric Wainwright (1996) argues that "a digital library remains a library, with the same purposes, functions and goals as a traditional library." Its key functions, therefore, are acquiring, organizing, storing, and providing access to digital resources. Peter Graham (1995) makes a similar point: "a Digital Research Library is a collection of electronic information organized for use in the long term" (p. 331). Its primary requirement "is that from the start it be committed to organizing, storing and providing electronic information for periods of time longer than human lives" (p. 332).

One assumption is that this coherent and unified collection of digitized materials is made available over the Internet or an Intranet. This can encompass a range of different media—text, image, sound, video —and of subject areas. The unifying factors are likely to include:

- coherent design and structure,
- consistent appearance and interface,
- library-wide standards, and
- the ability to search the digital library as a whole.

In some senses, the digital library represents the next stage of development beyond separate specialized centers or services offering electronic texts, digital images, and the like. Some universities are building their previous initiatives into a broader digital library. At the University of Michigan, for instance, the existing *Humanities Text Initiative*—one of the world's largest collections of electronic texts—has now become a component part of the university's Digital Library Production Service (Powell and Kerr, 1997).

Many institutions are working on some form of digital library activity, either by digitizing their own material, by purchasing material from commercial firms, or by developing an integrative framework for others' resources. Organizational approaches vary. In some universities, there is a distinct digital library group, which may also offer specialized training and support for users. In other contexts, the activities related to building the digital library are spread across the existing library organization, with each section taking responsibility for the tasks that relate most nearly to its existing role.

The University of California has prepared an ambitious plan for developing the California Digital Library. Organizationally, this will form another distinct library, within the distributed system of the university, under its own separate management. Five main roles are proposed for this library:

- The preservation, storage, and retrieval of information
- Access to, and delivery of, information through electronic networks
- On-line publishing of scholarly and scientific knowledge
- Consultation and training in the management of information
- Support for the knowledge network of the university

The content of such a library will be the first priority, over and above its technical features. It will provide access to electronic versions of existing publications, new electronic publications, specialized databases of images and textual corpora, and primary resources such as archives, manuscripts, and newspapers. These materials will be stored in a distributed network rather than a single repository. The first priority is to develop a Science, Technology and Industry Collection (University of California, 1996; Starr, 1998).

Elsewhere, digital libraries are being designed and implemented on a national scale. The Digital Library Initiative, funded by DARPA, NASA, and the National Science Foundation, is sponsoring six major digital library research projects in North America:

- University of Illinois at Urbana-Champaign: journal articles in computer science, electrical engineering, physics, civil engineering, and aerospace engineering
- University of California at Berkeley: materials relating to environmental planning, including photographs, satellite images, maps, and documents
- University of California at Santa Barbara: geographically referenced and spatially indexed maps and other materials (Project Alexandria)
- Stanford University: collections of computing literature, with an emphasis on designing a uniform way of accessing a variety of information sources
- Carnegie-Mellon University: a multimedia library with over 1,000 hours of digitized video, accompanied by audio, images, and text (the Informedia project)
- University of Michigan: earth and space sciences, with an emphasis on the use of software agents

The major goal of these projects is a technical one: to investigate the possibility of what is known as "deep semantic interoperability"—defined as the ability of a user to access similar classes of digital objects and services, in a distributed network of repositories, using mediating software to control the search process. The content of these digital library projects seems to have been chosen more for its ability to test technical aspects of their design than for its intrinsic value as a record of knowledge in particular disciplines. The National Science Foundation's briefing document for the second phase of the Digital Library Initiative (1999-2003) again emphasizes the methodological side of such projects, in areas such as "wide-spectrum information discovery," "intelligent user interfaces," "efficient data capture," interoperability, networked architectures, and "systems scalability." But users' needs are seen as the "driving motivation" of digital library research, and considerable attention is given to goals connected with improving education, learning, and scholarly communication and to developing repositories of knowledge on a global scale.

Although the use of the term "digital library" is perfectly justifiable in the context of the University of California's plan, its use on a wider scale seems to be somewhat misleading. The Library of Congress has assembled a Digital Library Federation, with fifteen research libraries and archives as members. Its aim is to build a structure for digitizing, preserving, and making accessible the historical materials relating to the cultural heritage of the United States. This envisages the creation of a distributed national digital library composed of two major strands:

- The conversion to digital versions of documents already held in the member libraries and archives
- The incorporation of existing electronic materials

At present, though, these exist as fifteen or more separate projects, with no interaction or integration between them. They are a "national digital library" only in the sense that the combined holdings of all the research libraries make up a "national library." Similarly, there has been talk of the "global digital library," and several conferences have been devoted to this theme. This term, too, tends to refer mainly to identifying common standards for individual digital libraries rather than designing and building an integrated global library.

Digital libraries are different from electronic text centers in their very nature, not just in the scale of their activities. A digital library exists as a networked service, not as a physical section within a library's buildings. It is an electronic parallel to the physical library, not a subsidiary part of it. One way of thinking of the relationship is by analogy with the transparent overlays attached to diagrams of the human body often found in encyclopedias. The digital library is an overlay on the existing physical library. The California Digital Library illustrates this well: it is a separate component of the University of California library system, operating on the same level as the libraries of the nine campuses.

The issues affecting the work of electronic text centers are even more important in the broader context of the digital library. Selection of materials is critical, as is arranging for their preservation.

Standards, particularly for data formats and for metadata, are also an integral part of a properly designed digital library. At present, however, much of the effort in digital library projects seems to be directed toward other ends, particularly the development of technical architectures and techniques for interoperability.

Commercial Services

Digital libraries, similar to electronic text centers, are largely a phenomenon of the academic environment. Despite their novelty of form and content, they still belong under the same umbrella as libraries and other support centers—services that exist to mediate between the academic community and the world of published texts and other resources. Although there has undoubtedly been some blurring of functions among libraries, computing centers, and academic departments, the outer boundary remains much the same.

A new development, with potentially profound implications for these relationships, is also now beginning to emerge. In this scenario, a commercial firm provides an electronic text service to researchers and students, bypassing the existing academic services to a considerable degree and taking on something of the library's role. This trend is occurring on several fronts, not just in the provision of scholarly electronic texts. Indexing and abstracting databases, reference works, and even electronic journals are now being offered in this way. The common—and novel—characteristic is that access to these resources is available, for an annual subscription, over the World Wide Web. Some publishers, such as Encyclopaedia Britannica, offer their products directly in this way, but more usually it is a third-party broker or agent that provides the service. Database providers such as Ovid, UMI, and Silver Platter, and subscription agents such as Blackwell's and Swets, are all marketing services of this kind.

The first, and best, example of this in the field of electronic texts is the *Literature Online* (or LION) service launched by Chadwyck-Healey Ltd. at the end of 1996 (see Figure 4.4). Advertised as "the home of English literature on the World Wide Web," *Literature Online* offers thousands of electronic texts of poetry, drama, and fiction in English. As well as these texts, there are reference materials—the *Annual Bibliography of English Language and Literature*

FIGURE 4.4. *Literature Online*, Chadwyck-Healey Ltd. (Home Page)

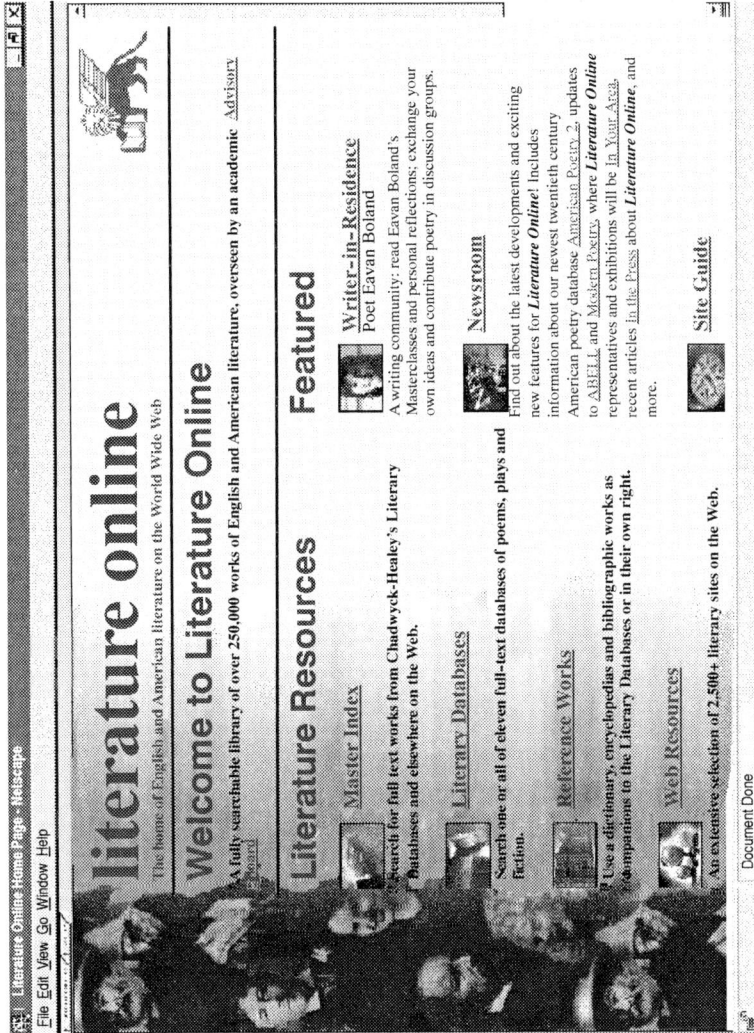

literature online

The home of English and American literature on the World Wide Web

Welcome to Literature Online

A fully searchable library of over 250,000 works of English and American literature, oversen by an academic Advisory Board.

Literature Resources

Master Index

Search for full text works from Chadwyck-Healey's Literary Databases and elsewhere on the Web.

Literary Databases

Search one or all of eleven full-text databases of poems, plays and fiction.

Reference Works

Use a dictionary, encyclopedias and bibliographic works as companions to the Literary Databases or in their own right.

Web Resources

An extensive selection of 2,500+ literary sites on the Web.

Featured

Writer-in-Residence
Poet Eavan Boland

A writing community: read Eavan Boland's Masterclasses and personal reflections; exchange your own ideas and contribute poetry in discussion groups.

Newsroom

Find out about the latest developments and exciting new features for *Literature Online*! Includes information about our newest twentieth century American poetry database American Poetry 2, updates to ABELL and Modern Poetry, where *Literature Online* representatives and exhibitions will be In Your Area, recent articles in the Press about *Literature Online*, and more.

Site Guide

and *Webster's Dictionary*—and an index to other sources of scholarly literary texts on the Internet. Parallel services from the same company offer other electronic text collections over the Web, including the *Patrologia Latina Database* and *Die Deutsche Lyrik*. In most cases, the resources are also published on CD-ROM and magnetic tape, but the Web version removes the need for local equipment and expertise beyond that needed to provide access to the World Wide Web.

Services such as *Literature Online* are aimed directly at the academic user. In the face of such commercial services, there are really only three major roles for the local academic library:

- *To act as a broker in arranging subscriptions and licenses.* It is still envisaged that libraries will be the subscribers to such services, although one can imagine that, in the future, users may be able to subscribe directly. But a library subscription has major advantages for both publishers and users, in terms of paperwork, discounts, and so on—just as with subscriptions to printed journals. In both Britain and Australia, an additional layer of discounts has been achieved by university libraries negotiating a national agreement with publishers and distributors of electronic materials.

- *To provide a suitable local gateway.* Institutions that are likely to subscribe to a service such as *Literature Online* will almost certainly have the necessary technical infrastructure for access to the World Wide Web. But the local library can build on this to provide a suitable intellectual framework for access to the service. At its simplest, this can be a link on the library's home page. At a higher level, this will involve a more extensive and coherent structure for Web resources that organizes them by subject and/or format. The University of Western Australia's CygNET Online service is one of many such frameworks, aimed at providing systematic Web access to a wide range of resources (see Figure 4.5). Without such a structure, the specific service will not be readily accessible and will only be available through personal bookmarks, knowledge of the direct URL, or a Web search.

- *To arrange suitable promotion and training for the service.*
  Web-based services are largely self-explanatory, but there will
  always be room for a focus of local expertise in their use. The
  library is well-placed to provide assistance and instruction in
  the use of the commercial services. Closely related is the need
  to promote the service; the cost of such a subscription—often
  quite expensive—justifies a considerable effort to ensure that
  potential users are aware of its availability.

There are at least three major drawbacks to a purely commercial
approach, however:

- Commercial publishers are only likely to publish or distribute
  material that will earn a reasonable return for them. They are
  not philanthropic institutions that will make any type of text
  available to researchers. There are always going to be resources which, for a variety of reasons, are unlikely to find a
  commercial publisher—particularly given the expense of commercial publication—but which are still of considerable interest and value to a range of researchers. Commercial electronic
  text services need to be supplemented by local university facilities for electronic publication and distribution.
- It is not clear, in many cases, whether libraries that subscribe
  to electronic services of this kind actually acquire any ownership rights as a result. When a subscription to a printed journal
  is canceled, the library retains the back set as its own property,
  for as long as it chooses. When an electronic subscription is
  canceled, the library may be left with nothing to show for its
  payments. This has been a feature of the subscription terms of
  several major publishers of CD-ROM databases, who have in-
  sisted that disks be returned after a subscription (or, more correctly, a leasing arrangement) is canceled. This is even more
  likely to occur when the subscription is not a physical product
  such as a CD-ROM, but a resource accessed over the World
  Wide Web. Johns Hopkins University Press has stipulated that
  subscribers to its Web-based electronic journal service, *Project
  Muse*, will receive a CD-ROM copy of the issues published in
  the subscription period. OCLC's *Electronic Journals Online*
  also promises continuing archival access to journal issues after

cancellations. However, these appear to be exceptional arrangements; most services do not offer any residual ownership after the subscription period has ceased.

- Most commercial publishers do not address the questions of archiving and preserving their data for the long term. There is no guarantee that the publications offered through these services will still be available in the future. If a publisher of printed books or journals goes out of business or stops publishing certain titles, the material that has already been published will still be available through libraries. But, with the Web-based subscription services arrangements that guarantee future access through an independent repository such as a library are unlikely. At present, it is possible to imagine a future scenario in which some of this material becomes entirely unavailable, despite the substantial investment in subscriptions by universities and their libraries.

Selection of materials is also an important issue for subscribers to commercial Web-based services. Subscribers to *Literature Online*, for example, are given the choice of a variety of packages containing different mixes of material at a range of prices. But many of the electronic journal packages are "all or nothing": the subscription must be to the entire package, and there is no opportunity to select individual titles. Once again, OCLC's *Electronic Journals Online* is an exception, in that libraries can subscribe to individual journals. But an OCLC access fee is also payable, and this is calculated on a sliding scale linked to the total number of subscriptions.

Bibliographic information—or metadata—is usually problematic, in that commercial services tend only to provide minimal author and title records which can be searched by keyword or browsed. Though it is not necessary to provide full Machine Readable Cataloging (MARC) records, there is no agreement on an alternative standard format for metadata, and no evidence that protostandards such as the Dublin Core are being used. Metadata records are not usually enhanced with additional subject descriptors or information on the bibliographic history of the material, such as earlier titles and editions. It is extremely difficult for libraries to integrate metadata

FIGURE 4.5. CygNET Online, The University of Western Australia Library

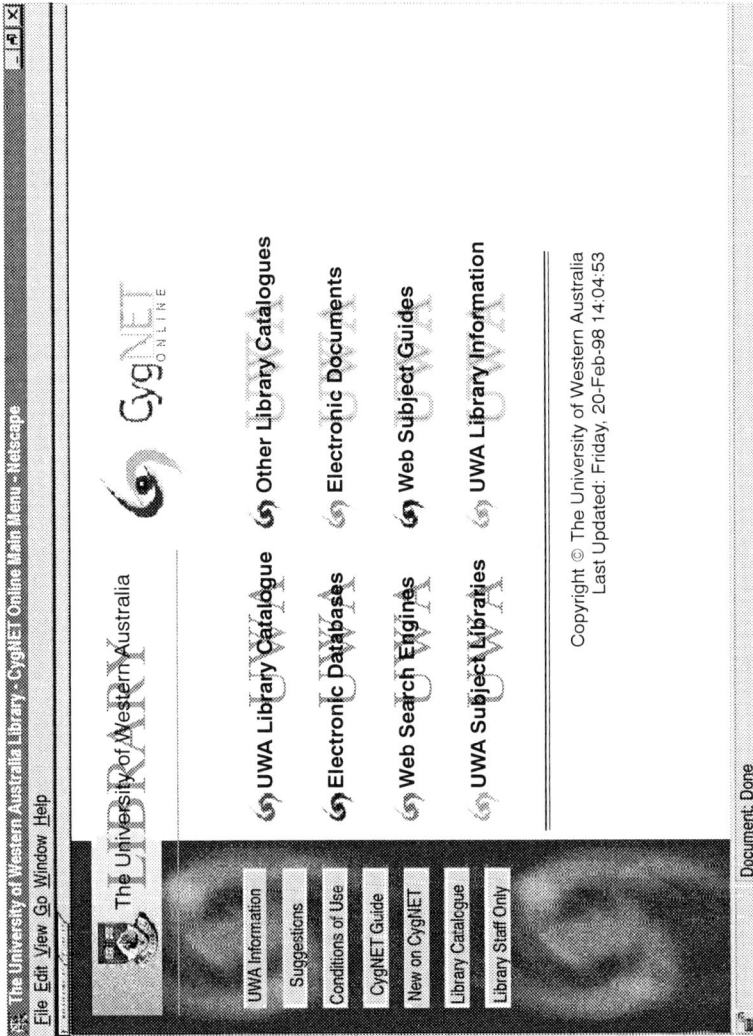

for this material into their on-line catalogs, except at the most summary level.

Publishers of electronic texts are likely to conform to standard formats for their resources. Both Chadwyck-Healey and Intelex, publishers of philosophical texts, use SGML-based markup for their products, which can be accessed with most standard Web-browsing software. Publishers and distributors of electronic journals, on the other hand, prefer to use more proprietary formats such as PDF and Real Page, which require additional viewing software such as Adobe Acrobat Reader and CatchWord.

The effect of these commercial services on the future development of electronic text centers and digital libraries is still largely unknown. But it seems reasonable to assume that universities and colleges which do not already have such local initiatives will be much less likely to try to develop their own in the future. After all, the cost of establishing and maintaining electronic text centers and digital libraries is considerable. Specialized expertise and knowledge are required. This is likely to be beyond the reach of all but the larger institutions. The cost of a subscription is likely to be more affordable and may bring some—if by no means all—the benefits of having networked electronic texts.

It remains to be seen whether the major issues relating to archiving, preservation, and ownership can be settled to the satisfaction of the academic and library communities. Even if they are, however, commercial arrangements should remain only one part of the way in which electronic texts are made available. They are no substitute for libraries or similar noncommercial services, which will be able to publish and distribute scholarly texts of only limited value to the commercial sector. Even more important, libraries will provide a local interface to the full range of networked services, tailored specifically to the needs of a particular institution and its researchers, teachers, and students.

National and International Services

Another important framework for the delivery of networked electronic texts is the national or international noncommercial service. Instead of individual institutions offering locally assembled packages for their own clientele, this involves national or international collaboration in developing a delivery mechanism. The usual means of support for such services is through targeted grants from government agencies or private foundations, but some services are also subsidized by individual universities as a service to the wider world of scholarship.

A particularly interesting case study of this kind is the Arts and Humanities Data Service (AHDS), established in Britain in 1996. Funded by a generous three-year grant from the Joint Information Systems Committee (JISC) of the Higher Education Funding Councils, the AHDS is intended to provide a national approach to creating, using, and accessing electronic resources in the humanities. Its role emphasizes collaboration and coordination between the users of these resources and the creators and providers of them, whether they are commercial or noncommercial. It also collects, catalogs, and preserves electronic resources, as well as providing support to users and promoting standards and guidelines for best practice in this field. The AHDS is, in many senses, a national digital library for the arts and humanities (see Figure 4.6).

In keeping with its emphasis on coordination and collaboration, the AHDS does not create and distribute electronic materials itself. Instead, it works through a distributed framework of service providers, which are associated with it through formal service agreements. These service providers are located in different parts of Great Britain and cover five main subject areas, or formats:

- The Oxford Text Archive (at Oxford University)
- The Historical Data Unit (at the University of Essex)
- The Performing Arts Data Service (at the University of Glasgow)
- The Archaeology Data Service (at the University of York)
- The Visual Arts Data Service (at the Surrey Institute of Art & Design)

The first two of these existed for many years before the advent of the AHDS, but have been successfuly incorporated into its framework. Each service provider is expected to provide various services: data collection, resource description and cataloging, on-line access, preservation, distribution, resource creation and enrichment, the development of standards, and training. The AHDS is currently working on

FIGURE 4.6. Arts and Humanities Data Service (Home Page)

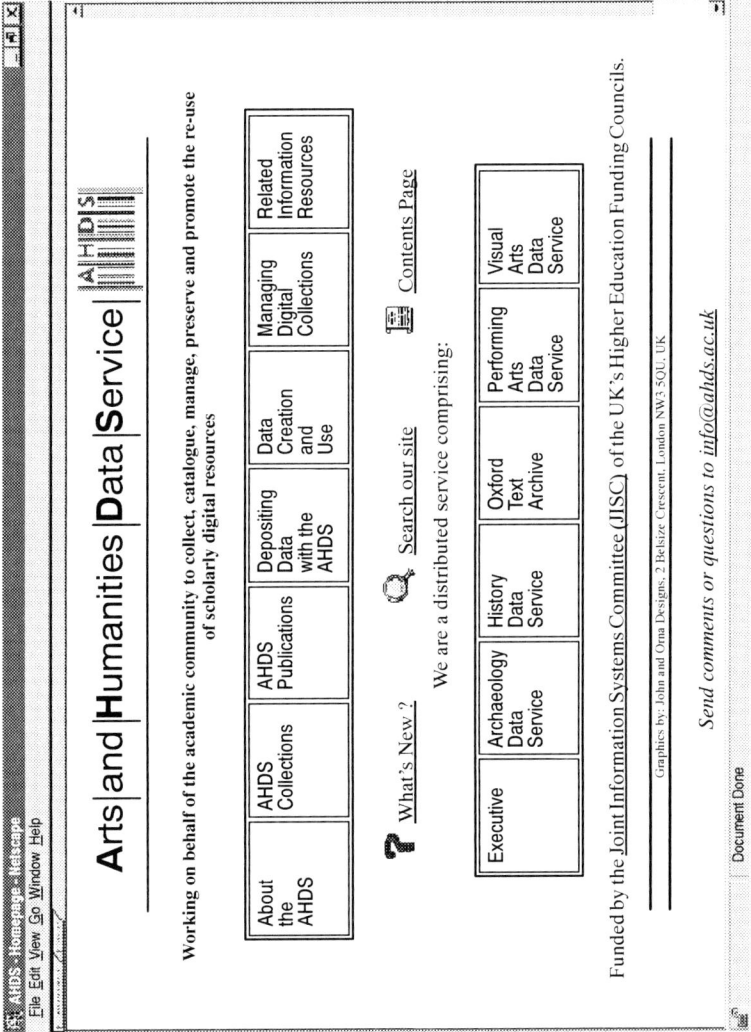

Arts and Humanities Data Service

Working on behalf of the academic community to collect, catalogue, manage, preserve and promote the re-use of scholarly digital resources

| About the AHDS | AHDS Collections | AHDS Publications | Depositing Data with the AHDS | Data Creation and Use | Managing Digital Collections | Related Information Resources |

What's New ?   Search our site   Contents Page

We are a distributed service comprising:

| Executive | Archaeology Data Service | History Data Service | Oxford Text Archive | Performing Arts Data Service | Visual Arts Data Service |

Graphics by: John and Orna Designs., 2 Belsize Crescent, London NW3 5QU, UK

Funded by the Joint Information Systems Committee (JISC) of the UK's Higher Education Funding Councils.

Send comments or questions to info@ahds.ac.uk

Document Done

ways of searching for resources across all the service providers simultaneously.

From the point of view of electronic texts, the Oxford Text Archive (OTA) is the most important of these service providers. Its mission is "to collect, catalogue, manage, and preserve digital text data of interest to textual scholars, and to distribute these data and encourage their scholarly re-use in research and teaching" (Oxford Text Archive, 1997). The OTA was set up in 1976 and now holds over 1,500 titles in forty different languages, covering a wide range of literary and linguistic areas of interest. These texts were originally distributed on diskette or magnetic tape but are now usually supplied by FTP (File Transfer Protocol) via the World Wide Web (see Figure 4.7). The immediate priorities of the OTA are to expand and consolidate its holdings, to enhance and improve bibliographic access to them, and to increase its training and promotional activities.

The OTA and the AHDS are an important development in the provision of access to electronic texts. The services they provide are, first and foremost, national, but their reach is global in practice. Mirror sites for the OTA, in particular, exist in North America and Australia. How the AHDS will continue after 1998, when its initial three-year grant is finished, remains to be seen. It may well adopt a subscription model aimed at covering its costs.

A different kind of approach, but with a similar aim, is offered by some of the so-called "virtual subject libraries" on the World Wide Web. These are usually based in a particular university or college and provide their services freely to the global Internet audience. Although they tend not to contain much in the way of textual material on their own sites, their value is in drawing together links to resources held all over the world. They provide integrated access to these resources by imposing a logical and consistent structure on them.

A good example of this kind of service is the *Labyrinth Library of Medieval Studies*. Based at Georgetown University, the *Labyrinth* provides high-level access to corpora of medieval texts at sites around the world, as well as links to sites covering a wide range of subjects (both geographically and topically) in medieval culture, resources for teaching, and information on professional organizations, societies, and publishers (see Figure 4.8). The textual section of the *Labyrinth* site, known as "auctores et fontes" (authors and

FIGURE 4.7. Oxford Text Archive (Home Page)

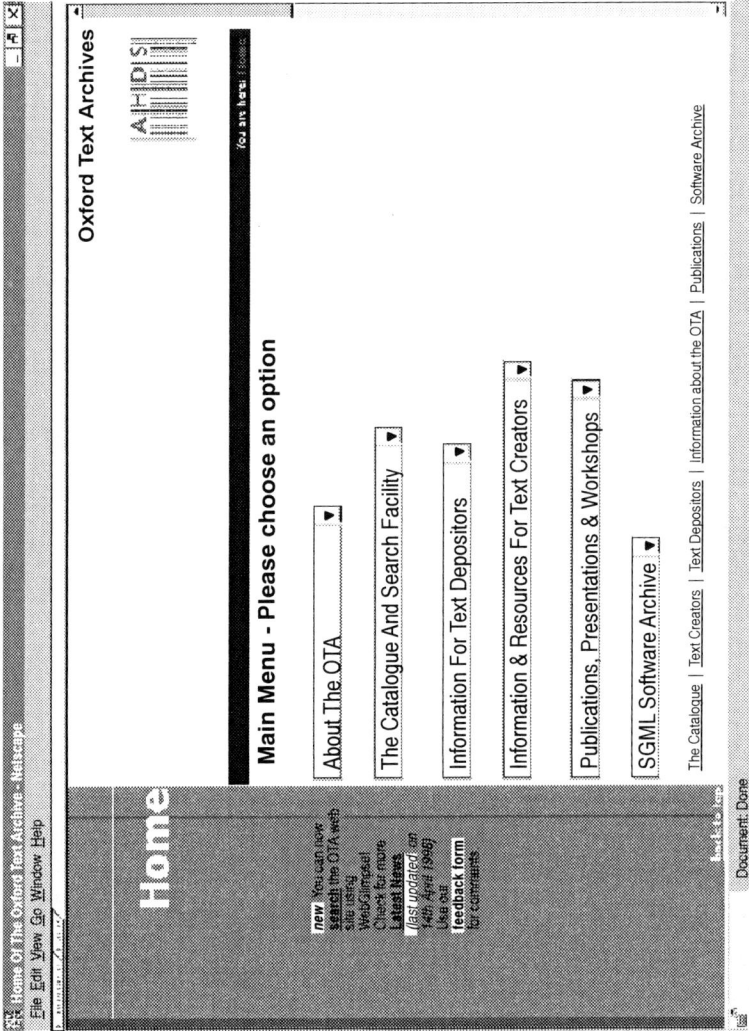

FIGURE 4.8. *Labyrinth Library of Medieval Studies* (Home Page—Detail)

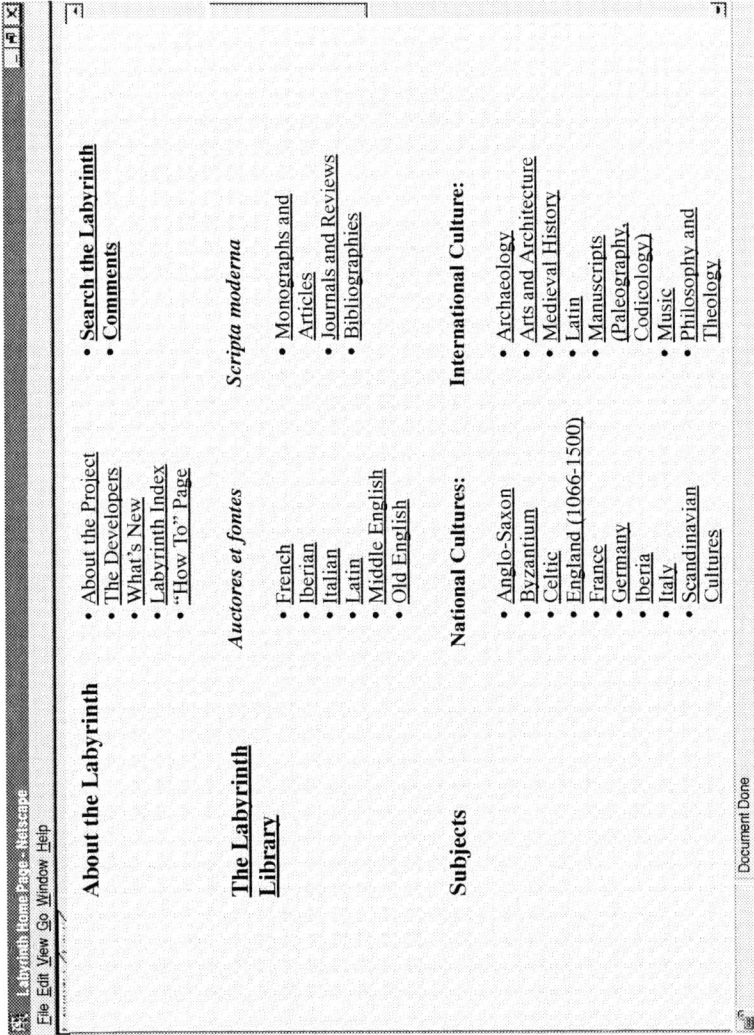

File Edit View Go Window Help

About the Labyrinth

- About the Project
- The Developers
- What's New
- Labyrinth Index
- "How To" Page

- Search the Labyrinth
- Comments

The Labyrinth Library

Auctores et fontes

- French
- Iberian
- Italian
- Latin
- Middle English
- Old English

Scripta moderna

- Monographs and Articles
- Journals and Reviews
- Bibliographies

Subjects

National Cultures:

- Anglo-Saxon
- Byzantium
- Celtic
- England (1066-1500)
- France
- Germany
- Iberia
- Italy
- Scandinavian Cultures

International Culture:

- Archaeology
- Arts and Architecture
- Medieval History
- Latin
- Manuscripts (Paleography, Codicology)
- Music
- Philosophy and Theology

Document Done

117

sources), is arranged by language: French, Iberian, Italian, Latin, Middle English, and Old English. The most extensive of these, the Latin section, lists more than seventy links to texts published on the Web at a wide range of sites in North America and Europe. Among them are James O'Donnell's individual versions of late Latin works. The *Labyrinth*'s Middle English texts, in contrast, derive almost completely from the collections of the University of Virginia's Electronic Text Center. The value added by the *Labyrinth* service is to place all these texts into a topical- or subject-based framework intended for a specific group of medievalist scholars and others with an interest in the Middle Ages.

Services such as the *Labyrinth* build on a distributed collection of textual resources and use the inherent characteristics of the World Wide Web to add an integrative layer across the top of this collection. The integration in this case is comparatively simple, consisting of browsable lists on Web pages. Accompanying these browsing services are specific search services such as Argos that have selected and vetted material for its scholarly content—unlike more general Web search engines such as Alta Vista and Lycos. Argos indexes the content of eleven scholarly sites in classics and medieval history and bills itself as the first "limited area search engine" on the Internet (see Figure 4.9). It also advertises its process of peer reviewing and accreditation of sites for searching. The "limited area" searching of Argos is a valuable counterpoint to the browsable lists of most scholarly sites.

Some attempts have been made to develop a globally available catalog of electronic texts. The Alex service was developed between 1993 and 1995 to explore methods of cataloging such texts, but lapsed for lack of funds and staff. It consisted of a database of about 2,000 entries that could be browsed or searched by keyword. Originally based on the Gopher protocol, it then became a Web database (known as Alcuin) with MARC records. A somewhat different approach is used by two current sites that aim to provide a global listing of electronic texts. The *Internet Public Library* contains links to about 6,500 texts or portions of texts. They can be browsed by author's name or by broad subject classification (based on the Dewey decimal system), and can also be searched by author, title, or Dewey subject. The *On-line Books Page* provides links to about 3,500 texts, limited to full books available freely over the

FIGURE 4.9. Argos (Home Page—Detail)

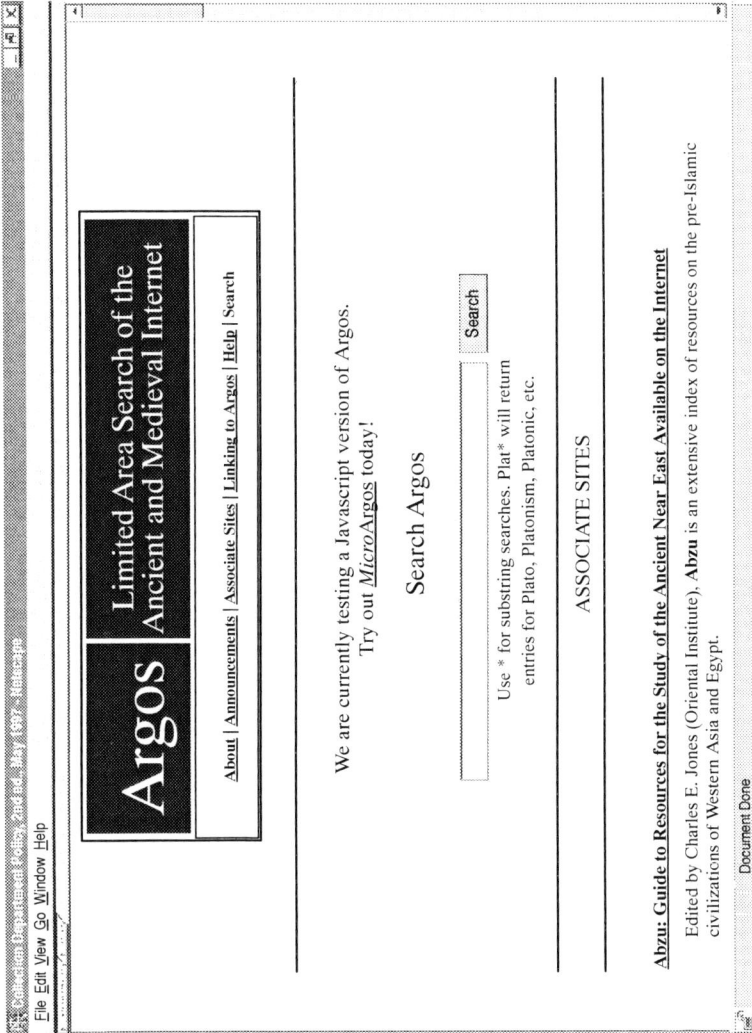

Internet. The listings can be browsed by author's name, title, or subject classification (based on the Library of Congress classification system), as well as being searchable by author or title.

GENERIC ISSUES: METADATA AND PRESERVATION

A distributed structure will be fundamental to the future development of access to electronic texts. Large national repositories of electronic texts are not envisaged. Instead, a wide variety of data providers—commercial, academic, consortia, even individuals—will make their resources available through the Web. From the perspective of the individual scholar, what will matter most is the framework through which they gain access to these resources. This is likely to be a combination of several layers: the local university library's information system, global subject gateways such as the *Labyrinth Library of Medieval Studies*, national or regional gateways such as the Arts and Humanities Data Service, and commercial services such as *Literature Online*. Each of these layers will aim to integrate the resources of the Web in a somewhat different way. A further strand of integration will be provided by scholarly searching services such as Argos, possibly supplemented by catalog services covering electronic texts.

The generic issues faced by all these repositories of electronic texts are similar to those facing libraries:

- On what basis should material be selected for inclusion in the repository's collections?
- What arrangements should be made for archiving and preserving texts, to ensure their continued use in the future?
- How can access best be provided to the widest range of users?
- What kind of cataloging and bibliographic description should be provided?
- What standards should be adopted to ensure consistency and compatibility?

Cataloging and bibliographic description, in particular, have been the object of much discussion. Rebecca Guenther has argued strongly for the continuing relevance of the MARC format used for library on-line catalogs (Guenther, 1996), but this view finds little support within the electronic text community. There is a widespread belief that MARC is unsatisfactory for documenting electronic texts (Gaynor, 1996). Its major weaknesses have been identified as:

- its inability to provide an effective structure for analytical, non-bibliographic information, such as the profile description, encoding description, and revision history prescribed by the TEI;
- its inability to show hierarchical bibliographic structures, such as interlinked records for items in complex collections of material with different levels of analysis; and
- its poor handling of information about different versions of the same text.

The inadequacies of the MARC format arise from its origins as an electronic version of the printed catalog card. Each MARC record, similar to each catalog card, describes a discrete bibliographic item. There is no ready method for linking these records to one another in a way that reflects hierarchical relationships that exist within collections of works. The MARC format presupposes a flat, single-tiered structure. As Michael Buckland observes, this approach breaks down in the integrated, interrelated world of computer networks: "the future catalog will have to be multitiered and flexible and adaptive in operation" (Buckland, 1994:C).

For electronic texts in the humanities, there are various possible alternatives to MARC. The Electronic Text Center at the University of Virginia has used the headers from TEI-encoded files to provide the basic bibliographic description of a text. These TEI headers have then been processed through a Perl program to convert them to MARC-like records, which can be reused for library catalogs (Seaman, 1996b; Gaynor, 1994). The Dublin Core set of metadata elements, first drafted in 1995 and intended particularly for authors to describe their own Web resources, is another alternative to MARC. Work has also been done on mapping between the Dublin Core's fifteen elements and the MARC format (Library of Congress, 1995). An SGML DTD has even been created for the MARC format (Davis, 1996).

A great deal of work is currently being done in this field, generally referred to as metadata and resource description. Several developments are likely to have some importance for the bibliographic description and identification of electronic texts:

- Web-based interfaces for on-line library catalogs are becoming commonplace, with provision for direct links between a catalog record and the networked resource it describes.
- The Z39.50 standard sets out a protocol for enabling common searching across different library catalogs and databases, despite different software and record formats (Lynch, 1997b).
- PICS (Platform for Internet Content Selection) contains technical specifications for a labeling system for HTML files, mainly intended for parents and teachers to control children's access to sites, but capable of being used for categorizing resources.
- RDF (Resource Description Framework) is a specification currently under development to provide an underlying infrastructure for exchanging metadata on the Web; it uses XML as the basis for its transfer syntax.
- MCF (Meta Content Framework) is another proposal that uses XML to develop a common data model and vocabulary for metadata on the Web.

Educom's Instructional Management Systems (IMS) project draws together several of these threads to specify a method for labeling educational resources on the Internet. The IMS Metadata Specification contains more than forty metadata fields, including those previously developed for the Dublin Core. These can be combined into four defined sets: base, item, tool, and module. The base set is mandatory and provides a generic description of an educational resource, while the other three sets are optional and are used to give more specific information about versions of that resource. IMS Metadata tags will be represented in an XML/RDF format and will be attached to Web pages, enabling more precise searching than is currently possible.

There is tremendous activity and considerable uncertainty in the field of metadata and resource description, and it is too early to tell which approaches will come to be accepted as international standards. A major consideration will be to integrate the catalogs, lists, and other metadata records maintained by a variety of repositories so that a user can carry out a search across a range of them simultaneously. The records do not need to be in the same format, as long as there is a mechanism for mapping them against a single search. One possible approach to this is being investigated by the AHDS, which is looking at integrating the different records of its component service providers by using a distributed cross-domain catalog, or resource discovery system, based on the Dublin Core metadata format (Miller and Greenstein, 1997).

This area is vitally important for the successful organization of access to electronic texts. Unless effective methods of identifying and searching for texts are available, the potential user will have great difficulty identifying and finding relevant material. One possible outcome in the medium term would be a distributed system with multiple sources of bibliographic information in a variety of different formats that are integrated into a kind of "virtual catalog."

An equally critical issue for electronic text repositories is the preservation of digital files in a usable form for long-term access. One of the key roles of libraries has always been to ensure that printed materials are preserved for future use. In pursuit of this goal, an enormous investment of money and time has been put into identifying items that are in danger of deteriorating to an unusable condition. Elaborate techniques have been developed for treating acidic paper, in particular, to prevent further decay. Stringent policies have also been implemented to ensure that rare and vulnerable items are used appropriately and even to create microform surrogates of them for day-to-day use.

The preservation of digital materials is far more uncertain. The physical media on which digital data are stored are inherently unstable. The average life of magnetic tapes has been estimated at about five years under normal conditions and about twenty years under ideal storage conditions, while the average life of CD-ROMs has been estimated at ten years under normal conditions (Council on Library and Information Resources, 1998). Add to this the rapidity with which software and hardware become obsolete, and the challenge to preserve digital materals becomes enormous. However, although the scale of the problem has been clearly documented, solutions to it are proving harder to identify and implement. The report of the Task Force on Archiving of Digital Information, commissioned by the Research Libraries Group and the Commission on Preservation and Access, calls for a national approach to the funding of preservation programs and to the development of policies in this area. It also recommends an extensive effort to test and identify the best techniques for preserving digital information. Regular migration of data from obsolescent formats and systems to newer ones is seen as the key to successful preservation, but this will require substantial funds and a continuing commitment from institutions that create or store digital materials. Without such action, however, important parts of our cultural heritage will be lost.

FURTHER READING

The volume of essays edited by Lawrence Dowler (1997), on the theme of the academic library as a gateway to knowledge, includes several that deal with the broader issues affecting the organization and delivery of electronic texts. Articles by Smith and Tibbo (1996) and Ellis (1996) focus specifically on the development of electronic text centers in American university libraries.

On the topic of digital libraries, the Web magazine *D-Lib* is a useful source of information, and its British counterpart *Ariadne* deals with developments in the United Kingdom. The International Federation of Library Associations and Institutions (IFLA) maintains a useful Web site that collects information about digital library resources and projects.

Information on metadata projects such as MCF, PICS, and RDF can be found at the World Wide Web Consortium's Web site and through the IFLA metadata Web pages.

Preservation of digital files is covered very thoroughly in the report of the Task Force on Archiving of Digital Information (1996) and in a report by the Arts and Humanities Data Service (Beagrie and Greenstein, 1998). The Web magazine *RLG DigiNews* is a good source of current information.

WEB SITES

Alex: http://www.lib.ncsu.edu/staff/morgan/alex/alex-index.html

Argos: http://argos.evansville.edu/

Ariadne: http://www.ariadne.ac.uk/

Arts & Humanities Data Service: http://www.ahds.ac.uk/

Digital Library Federation: http://lcweb.loc.gov/loc/ndlf/

Digital Library projects:
Berkeley: http://elib.cs.berkeley.edu/
Carnegie-Mellon: http://www.informedia.cs.cmu.edu/
Illinois: http://www.grainger.uiuc.edu.au/
Michigan: http://www.si.umich.edu/UMDL/
Santa Barbara: http://alexandria.sdc.ucsb.edu/
Stanford: http://www-diglib.stanford.edu/diglib/

D-Lib magazine: http://www.dlib.org/dlib/

Dublin Core: http://purl.oclc.org/metadata/dublin_core/main.html

IFLA—Digital Libraries: Metadata Resources:
http://ifla.inist.fr/II/metadata.htm

IFLA—Digital Libraries: Resources and Projects:
http://ifla.inist.fr/II/diglib.htm

IMS Metadata Project:
http://www/imsproject.org/md_overview.html

Internet Public Library: http://www.ipl.org/

James J. O'Donnell: http://ccat.sas.upenn.edu/jod/

Labyrinth Library of Medieval Studies:
http://www.georgetown.edu/labyrinth/labyrinth-home.html

MCF: http://www.w3.org/TR-NOTE-MCF-XML/

National Science Foundation, Digital Library Initiative, Phase 2:
http://www.nsf.gov/pubs/1998/nsf9863/nsf9863.txt

On-Line Books Page: http://www.cs.cmu.edu/books.html

Oxford Text Archive: http://ota.ahds.ac.uk/

PICS: http://www.w3.org/PICS/

RDF: http://www.w3.org/RDF/Overview.html

RLG DigiNews: http://www.rlg.org/preserv/diginews/

Chapter 5

Structures, Architectures, and Editions

The format of the printed book has been developed and refined over more than 400 years, and is now a fixed and familiar cultural artifact. Electronic texts are very different. An electronic text can be presented to its users in a wide variety of ways, leaving room for considerable experimentation and variation in its structure, organization, and format. Many different approaches can be employed, ranging from a simple, linear presentation to an elaborate network of interlocking textual materials. The editor or publisher of an electronic text can even choose to present more than one of these structures in parallel.

The relationship between the electronic medium and scholarly editing is a complex one. Editing is a discipline with a history of considerable debate and disagreement over appropriate approaches to adopt in preparing and presenting an edition of a text. Electronic texts draw on the various views encompassed by this tradition, as well as adding some new choices and alternatives in textual editing. For some commentators, the electronic medium has a natural affinity with postmodernist, "reader-centered" texts rather than with the typical critical edition in printed form.

In this chapter, we look at the types of structures and architectures that can be applied to electronic texts and also consider the relationship between electronic texts and the different types of editions that can be produced.

STRUCTURE AND ARCHITECTURE

Among the most important characteristics of electronic texts are their structure and architecture. We do not mean by this the technical design of an electronic text, in the sense of the software and hardware it employs. Instead, the terms refer to the intellectual framework into which the text is arranged for presentation to the user. The conventional printed book has a standard and static structure, developed and refined over more than five centuries (Steinberg, 1974). It usually consists of a title page, preface, introduction, and table of contents, followed by the individual chapters or sections in sequence, with a bibliography and indexes to round out the volume. The text can be opened and read at any page but is normally intended to be read in a linear sequence, starting at the beginning and finishing at the end.

Much of the power of the printed book arises from the relationship between this intellectual framework and the physical structure of the volume. The two aspects mesh in a remarkable synergy, with each complementing the other. The physical aspect of the printed book is much older than printing itself and derives, ultimately, from the manuscript codex, which seems to have first emerged in the second century A.D. as an alternative to the scroll. Eric Turner, who wrote the definitive account of the early codex, places its unknown inventor among the "greatest benefactors of mankind" (Turner, 1977:1). After 2,000 years of refinement, the physical form of the codex-cum-printed book is a masterpiece of design, with every aspect working in harmony with the intellectual structure of the contents (Williamson, 1983). There is little apparent room—or reason—to experiment with this basic structure.

Nevertheless, the printed book does promote a particular way of writing and reading texts—at least according to its critics. It "encourages the notion of a text as an organic whole" and as a separate, distinct, and unique work, according to Jay David Bolter (1991:163). It promotes a linear approach in both exposition and reading. It encourages the assumption that the reader is a largely passive recipient of the author's message. For those who are keen to promote a different kind of reading and authorship, the success and stability of the printed book present major obstacles.

From the point of view of scholars working with texts in multiple versions, the printed book is also seriously inadequate. The critical edition, in its definitive printed form, is "infamously difficult to read and use," according to Jerome McGann (1995). This, in his view, is because the codex format is being used to present a study of something already in codex format. The solution to this problem generally involves an elaborate critical apparatus with a forest of bibliographical footnotes, shorthand references, and notes on variant readings, all very complicated to navigate and interpret. Also lost is the full richness of the editor's study of a text because of the difficulty—and often impossibility—of reproducing within the confines of a printed edition other materials related to the text, in both print and other media. A further inadequacy is that the printed book contains the most minimal analytical tools—usually consisting only of an index. Readers find it very difficult to carry out any more sophisticated analyses of the text.

In comparison with the printed book, electronic texts have an extremely short history. The earliest work on computerizing texts is generally considered to have been carried out by Fr. Roberto Busa, who prepared a concordance to the works of St. Thomas Aquinas in the 1940s (Hockey, 1996:1). Only in the 1990s have electronic texts become relatively commonplace. But even in that short time, a number of different structures have been tried, and various experiments with different architectures have been carried out. None of these has achieved universal acceptance, and no dominant structure, intellectual or physical, has yet emerged. It seems, at this stage, that variation and eclecticism may well be inherent in electronic texts. The experimentation characteristic of the current situation seems likely to last for the foreseeable future.

To what extent are different structures and architectures for electronic texts conditioned by the type of software being used to create them? It is certainly true that the structural choices were very limited for early work on electronic texts because the range of software was so small. In the 1990s, however, such a variety of different software is available for creating and publishing electronic texts that almost any structure can be created. Instead of the structure being dictated by the software, as tended to be the case in earlier years, software can now be chosen to achieve the desired structure.

Linear Presentation

At the lowest level, an electronic text can be presented as a single linear structure. To use it, the reader must call up the entire text and

scroll through it from beginning to end, usually on several successive screens. There is no alternative way of reading the text. Some
limited analytical capacity may be present, inasmuch as most word-
processing or Web-browsing software contains a simple method of
searching for a string of characters in a text file. In this context, the
use of the word "scrolling" to describe the process of reading such a
text is very revealing. This kind of structure is indeed strongly
reminiscent of the manuscript scroll of the classical world and
shares its disadvantages. If anything, an electronic scroll is even
more unsuited to the presentation of a text than a handwritten scroll
was. Very few people are comfortable with reading a lengthy text
on a computer screen in this way. The obvious solution is to print
out a copy on single sheets of paper, thus transforming it into the
codex form again.

Most "plain ASCII" texts are presented as a simple, linear structure, especially when they are delivered over the Internet by FTP as
a single file. Typical of this approach is *Project Gutenberg*, which
offers each text in a single file for downloading by FTP and viewing
with word-processing or text-editing software. This approach is
probably necessary, though not inevitable, for ASCII texts. The
only alternative is to break up a single lengthy text into multiple
separate files cannot be linked in any way. Most producers of such
texts have preferred to present their text as a single file to retain its
unity, even when it is very long.

Quite a few shorter HTML and SGML texts are also provided
over the World Wide Web as a single linear file. This is no doubt
based on the assumption that they are too short to make a segmented table of contents worthwhile. The shorter texts offered by
the University of Virginia's Electronic Text Center, for instance,
tend to be single linear files. The cutoff point appears to be 100 KB,
hence Chaucer's *Canterbury Tales* has a table of contents and separate files, but his *Book of the Duchess* does not (see Figure 5.1).
Similarly, Louisa May Alcott's *Little Women* has a table of contents
but her story "Scarlet Stockings" does not. The classical and medieval texts to be found in James O'Donnell's collection show a
similar variation. The *Fables* of Avianus appears as a single file,
while Augustine's *Confessions* has a table of contents and multiple

FIGURE 5.1. Electronic Text Center, University of Virginia Library: Middle English Collection (Partial Listing)

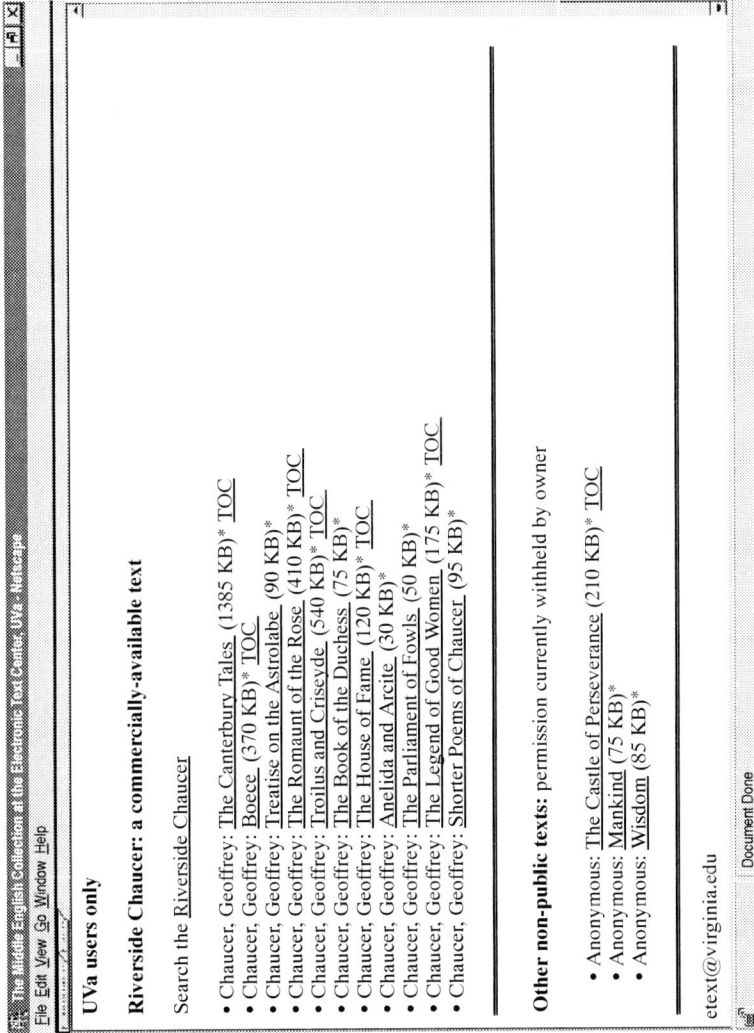

The Middle English Collection at the Electronic Text Center, UVa - Netscape

File Edit View Go Window Help

UVa users only

Riverside Chaucer: a commercially-available text

Search the Riverside Chaucer

- Chaucer, Geoffrey: The Canterbury Tales (1385 KB)* TOC
- Chaucer, Geoffrey: Boece (370 KB)* TOC
- Chaucer, Geoffrey: Treatise on the Astrolabe (90 KB)*
- Chaucer, Geoffrey: The Romaunt of the Rose (410 KB)* TOC
- Chaucer, Geoffrey: Troilus and Criseyde (540 KB)* TOC
- Chaucer, Geoffrey: The Book of the Duchess (75 KB)*
- Chaucer, Geoffrey: The House of Fame (120 KB)* TOC
- Chaucer, Geoffrey: Anelida and Arcite (30 KB)*
- Chaucer, Geoffrey: The Parliament of Fowls (50 KB)*
- Chaucer, Geoffrey: The Legend of Good Women (175 KB)* TOC
- Chaucer, Geoffrey: Shorter Poems of Chaucer (95 KB)*

Other non-public texts: permission currently withheld by owner

- Anonymous: The Castle of Perseverance (210 KB)* TOC
- Anonymous: Mankind (75 KB)*
- Anonymous: Wisdom (85 KB)*

etext@virginia.edu

Document Done

files. The English translation of Boethius's *Consolation of Philosophy* is a single file, while the Latin original is not.

Segmented Text

A somewhat greater level of sophistication is evident when an electronic text is divided into discrete segments, which are analogous to chapters in a printed book. A table of contents is usually present, hyperlinked to the individual segments. These can then be browsed individually, and in any order. With this kind of text, the reader can choose to go directly to a particular chapter or section and read from there and can read the complete text in any sequence. This has some similarity to a printed book, inasmuch as a book can be opened at any chapter and can be read out of its linear sequence.

This kind of presentation is common in HTML texts distributed on the Web—especially for large texts. A good example is William Schipper's HTML version of a large medieval Latin encyclopedia, composed by Rabanus Maurus in the early ninth century and known as *De rerum naturis* (see Figure 5.2). This electronic text is divided into twenty-four segments, representing the two prefaces and twenty-two books of this work. Each book is contained in a separate HTML file, with its own brief table of contents to enable the reader to jump directly to any of the chapters within the book. Navigation between the different books of this substantial work is made easier by the use of a separate "frame" for the main list of contents. This frame can be seen from anywhere within the text, with no need to return to the main contents screen.

The same approach is also used for some electronic journals. Those of the Johns Hopkins University Press, distributed on the Web under the name *Project Muse*, present each journal as a hierarchical series of HTML files (see Figure 5.3). The top file contains a list of individual issues, each of which is linked to a file containing a table of contents for that issue. This file, in its turn, is linked to an individual HTML file giving the complete text of each article. Articles can be read in any order the reader chooses.

An interesting effect of this kind of segmented structure is that it is difficult to conceptualize the whole text in the same way as one can grasp the totality of a book at a glance. In a segmented electronic text, the table of contents is the only overview available to the

FIGURE 5.2. William Schipper: Rabanus Maurus, *De rerum naturis* (Home Page)

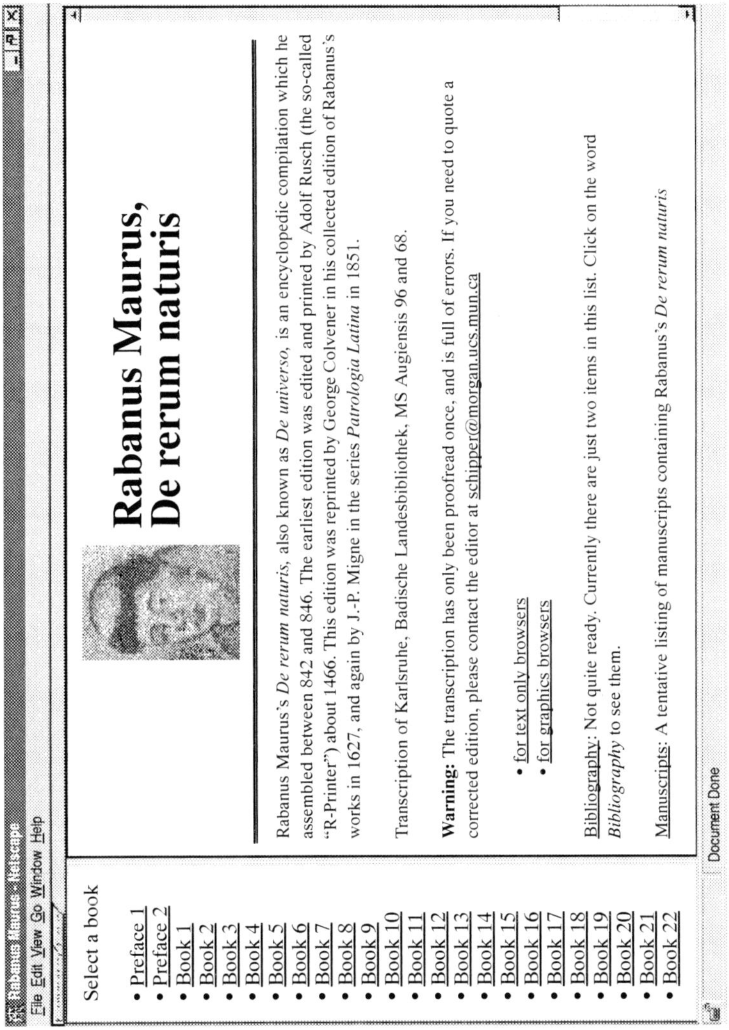
Rabanus Maurus - Netscape
File Edit View Go Window Help
Select a book
Preface 1
Preface 2
Book 1
Book 2
Book 3
Book 4
Book 5
Book 6
Book 7
Book 8
Book 9
Book 10
Book 11
Book 12
Book 13
Book 14
Book 15
Book 16
Book 17
Book 18
Book 19
Book 20
Book 21
Book 22
Rabanus Maurus, De rerum naturis
Rabanus Maurus's De rerum naturis, also known as De universo, is an encyclopedic compilation which he assembled between 842 and 846. The earliest edition was edited and printed by Adolf Rusch (the so-called "R-Printer") about 1466. This edition was reprinted by George Colvener in his collected edition of Rabanus's works in 1627, and again by J.-P. Migne in the series Patrologia Latina in 1851.
Transcription of Karlsruhe, Badische Landesbibliothek, MS Augiensis 96 and 68.
Warning: The transcription has only been proofread once, and is full of errors. If you need to quote a corrected edition, please contact the editor at schipper@morgan.ucs.mun.ca
for text only browsers
for graphics browsers
Bibliography: Not quite ready. Currently there are just two items in this list. Click on the word Bibliography to see them.
Manuscripts: A tentative listing of manuscripts containing Rabanus's De rerum naturis
Document Done

FIGURE 5.3. *Project Muse:* Johns Hopkins University Press (Sample Issue)

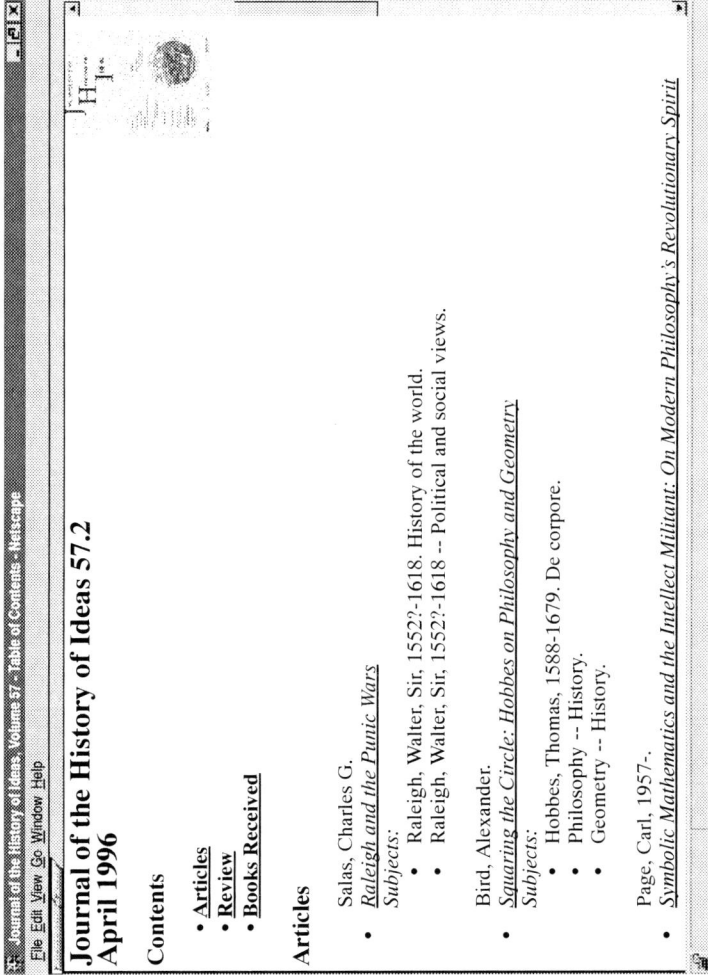

File Edit View Go Window Help

Journal of the History of Ideas 57.2
April 1996

Contents

- **Articles**
- **Review**
- **Books Received**

Articles

- Salas, Charles G.
  Raleigh and the Punic Wars
  Subjects:
  - Raleigh, Walter, Sir, 1552?-1618. History of the world.
  - Raleigh, Walter, Sir, 1552?-1618 -- Political and social views.

- Bird, Alexander.
  Squaring the Circle: Hobbes on Philosophy and Geometry
  Subjects:
  - Hobbes, Thomas, 1588-1679. De corpore.
  - Philosophy -- History.
  - Geometry -- History.

- Page, Carl, 1957-.
  Symbolic Mathematics and the Intellect Militant: On Modern Philosophy's Revolutionary Spirit

reader. Each segment must then be summoned separately to the screen in full. This does not seem to matter greatly when the text is, say, an issue of a journal or a compilation of conference papers. But, when the text is a coherent literary work, the effect can be quite unsettling and frustrating. No doubt this effect is a legacy from our familiarity with printed books and is less likely to worry future generations who are more accustomed to the electronic approach.

Image Files

A generally similar structure can be found in an electronic text that consists of a linked series of image files. These, too, usually have a table of contents and can be viewed by taking the files in a variety of possible orders. Again, there is a superficial resemblance to the printed codex, which is hardly surprising given that the images are usually of printed pages. But this format also shares with HTML files the difficulty in grasping the whole work at a glance. Depending on the type of image file used, this effect may be heightened by the lengthy time required to call up individual segments onto the screen. To some extent, these drawbacks can be reduced by techniques that group the page images for a chapter or an article into a single file. Both the PDF and TIFF formats can support these multipage files.

Many electronic journals are organized as a series of image files, most commonly in PDF or the Real Page Catchword format. The journals published through the IDEAL project of Academic Press, for instance, can each be browsed hierarchically, beginning with an HTML list of available years (see Figure 5.4). The HTML page for each year contains a list of issues, and each issue contains an HTML table of contents. This then reveals a list of articles for the issue, each with two choices: an abstract (in a separate HTML file) and the full text of the article. This full text consists of a series of linked PDF images that can be browsed with Adobe Acrobat software. These are, in fact, merely digital images of the printed pages of the journal.

One of the best examples of the application of image files to structuring electronic texts in the humanities can be found in the *Making of America* project, housed at the University of Michigan and Cornell University (see Figure 5.5). This collection of primary sources for nineteenth-century American social history already con-

FIGURE 5.4. *IDEALibrary:* Academic Press (Sample Issue)

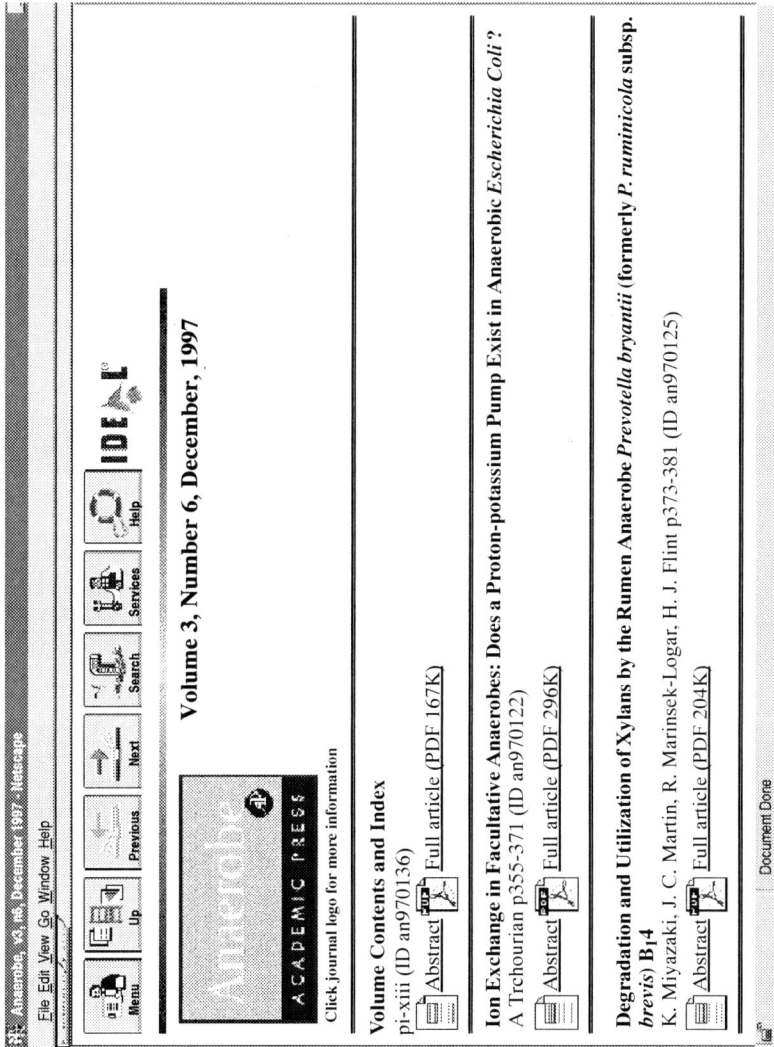

Anaerobe, v3, n6, December 1997 - Netscape

File Edit View Go Window Help

Menu | Up | Previous | Next | Search | Services | Help | IDE∧L

Click journal logo for more information

Volume 3, Number 6, December, 1997

Volume Contents and Index

pi-xiii (ID an970136)

Abstract | Full article (PDF 167K)

Ion Exchange in Facultative Anaerobes: Does a Proton-potassium Pump Exist in Anaerobic Escherichia *Coli*?

A Trchourian p355-371 (ID an970122)

Abstract | Full article (PDF 296K)

Degradation and Utilization of Xylans by the Rumen Anaerobe *Prevotella bryantii* (formerly *P. ruminicola* subsp. *brevis*) B₁4

K. Miyazaki, J. C. Martin, R. Marinsek-Logar, H. J. Flint p373-381 (ID an970125)

Abstract | Full article (PDF 204K)

Document Done

FIGURE 5.5. *Making of America* (Sample Issue)

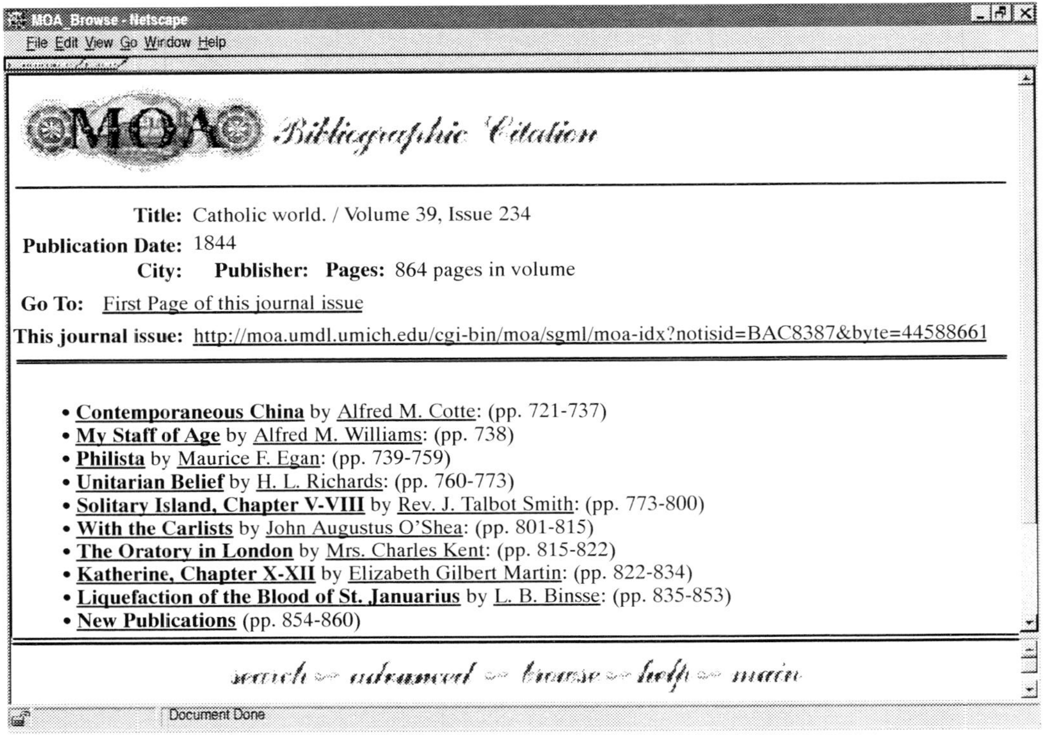
MOA_Browse - Netscape
File Edit View Go Window Help
MOA Bibliographic Citation
Title: Catholic world. / Volume 39, Issue 234
Publication Date: 1844
City: Publisher: Pages: 864 pages in volume
Go To: First Page of this journal issue
This journal issue: http://moa.umdl.umich.edu/cgi-bin/moa/sgml/moa-idx?notisid=BAC8387&byte=44588661
• Contemporaneous China by Alfred M. Cotte: (pp. 721-737)
• My Staff of Age by Alfred M. Williams: (pp. 738)
• Philista by Maurice F. Egan: (pp. 739-759)
• Unitarian Belief by H. L. Richards: (pp. 760-773)
• Solitary Island, Chapter V-VIII by Rev. J. Talbot Smith: (pp. 773-800)
• With the Carlists by John Augustus O'Shea: (pp. 801-815)
• The Oratory in London by Mrs. Charles Kent: (pp. 815-822)
• Katherine, Chapter X-XII by Elizabeth Gilbert Martin: (pp. 822-834)
• Liquefaction of the Blood of St. Januarius by L. B. Binsse: (pp. 835-853)
• New Publications (pp. 854-860)
search ~ advanced ~ browse ~ help ~ main
Document Done

tains over 1,600 books and 30 journals—over 4,000 volumes in all. Each of the 634,000 pages is presented as an image file, originally created in the TIFF format and compressed using the CCITT Group 4 standard. The page images are not grouped into a single file for each article, however. An individual file is used for each page, and the table of contents lists each page separately.

The sample reproductions of Australian colonial newspapers provided by the *Australian Cooperative Digitisation Project* also illustrate this kind of approach (see Figure 5.6). In addition, they offer the ability to compare the same printed material reproduced in different image formats: GIF, TIFF compressed using CCITT Group 4, and PDF. Although the reproduction of these formats varies, the overall structure of this electronic version remains much the same, regardless of the image format used. The only method of reading the text is to browse through the image files, which are arranged in the same order as the printed text. The electronic edition may also be designed to allow any specific page number to be viewed first, just as a printed book can be opened at any page. In some cases, pages are grouped into a multipage image file.

A somewhat different approach to the problem of grouping page images into articles or chapters is offered by the *Electronic Binding Project*, or *Ebind*. This uses an SGML Document Type Definition to provide the structure rather than relying on structural mechanisms in the image formats themselves. As applied in the *Digital Photocopy* and *American Heritage Virtual Archive* projects at the University of California Berkeley, an *Ebind* document consists of a series of image files that are linked together by an overarching SGML file. This control file records the structural hierarchy of the document— chapters and sections—and its printed pagination, as well as bibliographic information, access points, and abstracts. Links to the corresponding image files are embedded at the appropriate places in the structure. A Perl script then translates this file into a form that can be browsed through the World Wide Web.

Hypertext Structures

Textual structures that are either segmented or linear are still comparatively close to their printed predecessors. In fact, they tend to add to the limitations of the printed book rather than addressing

FIGURE 5.6. *Australian Cooperative Digitisation Project* (Sample Issues)

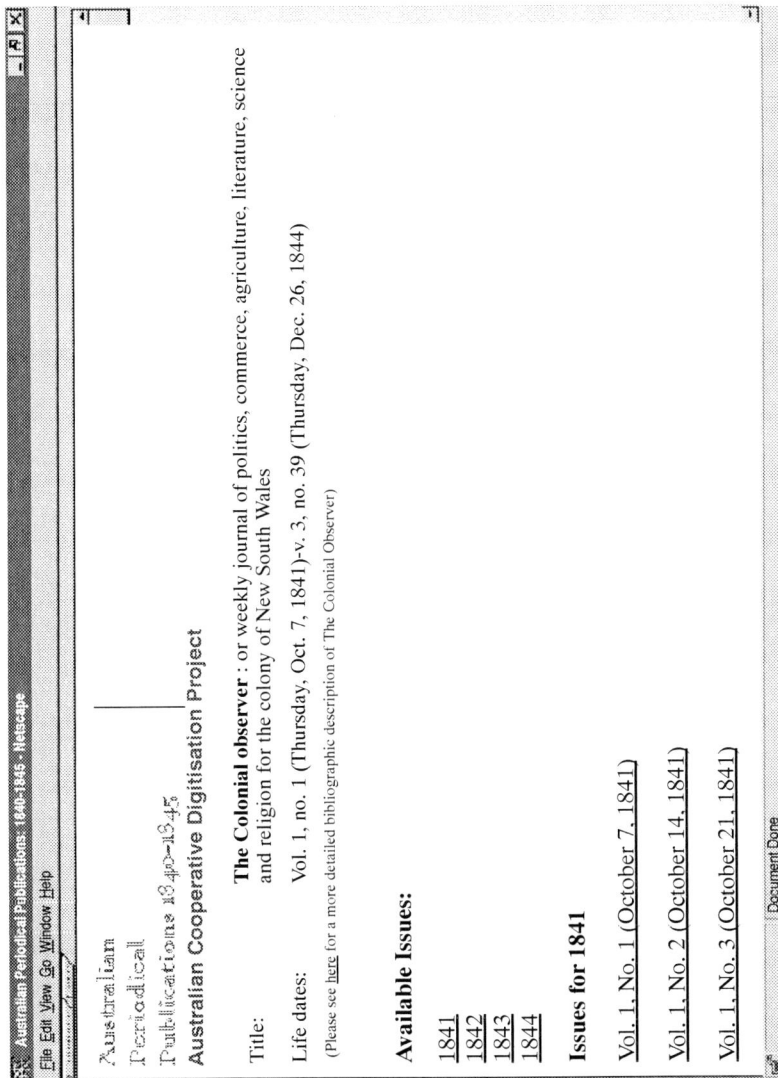

Australian
Periodical
Publications 1840-1845

Australian Cooperative Digitisation Project

Title: **The Colonial observer** : or weekly journal of politics, commerce, agriculture, literature, science and religion for the colony of New South Wales

Life dates: Vol. 1, no. 1 (Thursday, Oct. 7, 1841)-v. 3, no. 39 (Thursday, Dec. 26, 1844)

(Please see here for a more detailed bibliographic description of The Colonial Observer)

Available Issues:

1841
1842
1843
1844

Issues for 1841

Vol. 1, No. 1 (October 7, 1841)

Vol. 1, No. 2 (October 14, 1841)

Vol. 1, No. 3 (October 21, 1841)

Document Done

them. A distinctively electronic approach only begins to emerge when a text is presented in a hypertext structure. The term "hypertext" tends to be used very loosely and is sometimes applied to anything with HTML coding. It was defined by its originator, Ted Nelson, as "nonsequential writing" (Nelson, 1987:17). This definition is not particularly helpful if understood in a purely formal sense. After all, the form of the printed book has its own nonsequential aspects, notably the index and the ability to open at any page and start reading. Footnotes can also be considered as a hypertextual device. From this point of view, printed books are not so far removed from the simpler kind of electronic text with its table of contents linked to sections or chapters.

If Nelson's definition is taken to refer to the intellectual process of composition, however, its meaning becomes clearer. A hypertext is something that is designed and intended to be read in a nonlinear fashion. The text as a whole, and the individual textual fragments that constitute it, contain within them numerous embedded direct links to other materials. These links can be to related pieces of text, footnotes, or associated images and other kinds of media. They can also be links to analytical tools, enabling readers to ask their own questions of the text. They can take the reader to a greater level of detail and to smaller fragments of related material or to larger and broader structures within which the text is absorbed or located.

For the most part, this hypertextual structure reflects the views and ideas of its creator and designer. Someone has to decide what will link to what. The analytical tools can only work within the parameters specified by the person constructing the hypertext. Sooner or later, the reader of a hypertext reaches the limit of the different paths that can be taken through the weblike network of the text. In some (perhaps even many) publications—especially those of a popular, commercial nature—these limits are reached fairly quickly. The art of effective hypertextual design consists, therefore, in building as many interconnecting links and paths as possible rather than just those which the designer regards as valid or useful.

Even the most complex hypertext requires some kind of entry point, which is the rough equivalent of a table of contents. At one extreme, this may be a list of all the main units of the text, though these may not have the logical order we would expect of a table of contents. At the other, there may be only one starting point, which branches out into an interlocking tree structure at subsequent levels. In either case, however, the underlying structure is likely to resemble a network or web of textual fragments or components in other media, which can be navigated by a number of different routes.

The World Wide Web itself, from one point of view, can be considered a gigantic hypertext. There are a vast number of possible starting points, given that almost any Web page can be set quite easily as the "home page" for an individual user's browser software. These serve as a kind of table of contents, which is more or less comprehensive and effective, depending on its design and scope. The number of different paths through the Web is finite, at least theoretically, but seems infinite to the ordinary user. In this sense, the Web is the electronic equivalent of the infinite *Library of Babel* imagined by Jorge Luis Borges (1965:72-80).

Sections within the Web also form more coherent hypertexts. The texts listed by the Internet Public Library, for example, amount to a kind of metatext composed of 6,000 individual parts. The *Labyrinth Library of Medieval Studies* is another example of this structure, since it brings together numerous individual medieval texts arranged in a loose hypertext network. The links provided by a service such as the *Labyrinth* are comparatively few and simple and consist mainly of lists that bring together texts with similar characteristics—language, subject, and so on. There is nothing in the way of an analytical capability.

At the level of the single text, the thorough-going hypertext in Nelson's sense may be achieved by two means: creating extensive cross-references within an electronic version of a single text or assembling a range of different versions of a text, together with related material of various kinds, and building links between them. In both cases, the actual extent of the hypertext is determined by the designer or creator. The user is able to follow as many different paths through and around the text as its creator has allowed for. In the electronic version of Chaucer's *Prologue* to *The Wife of Bath's Tale*, for instance, there are around two million hypertext links within the text and an almost infinite number of paths through it.

For most of the electronic texts in the humanities currently available, these links and paths tend to be dominated by the limitations of the printed editions on which they are based. The most obvious internal linkages are between the main text and various ancillary materials: explanatory notes, footnotes, and bibliographies. The Chadwyck-Healey texts reproduce these annotations from the printed originals very faithfully. In the printed *Patrologia Latina*, for instance, there are multiple layers of commentary and footnoting preserved from the seventeenth- and eighteenth-century editions used by Migne. These can be followed as links in the electronic edition. In a similar way, the *Bible in English* contains internal cross-references between the biblical text and the various citations, notes, tables, and lists that appear in the original printed volumes. The technique used is for each of these to appear in its own small file, accessible from the main text by a suitable icon or number. In fact, all the material appears in the same electronic file; the markup and the presentational software between them contrive to make the notes appear as a separate file.

A different approach to hyperlinks can be found in the Ancient Greek literary works provided by the *Perseus Project* (see Figure 5.7). From each word of the Greek text, the reader can invoke a morphological analysis of that word. This provides, among other things, an explanation of the full form from which the word derives and a frequency count for that word in the specific author and in the *Perseus* textual corpus as a whole. Further links are also available, to the entry for the word in the standard dictionaries of Classical Greek and to all the occurrences of that word in the author being searched. It is also possible to see all the occurrences of the word in all other authors in the corpus. All these links and lists are presented as HTML pages, although they are, in fact, generated from an elaborate database. The sophistication of this approach reveals very clearly the power of well-designed hypertextual structures for the study of texts. A similar technique is being applied to the Classical Latin texts that the *Perseus Project* is beginning to make available. Further elaboration of this interlinking is planned, with bidirectional connections to commentaries and other secondary works.

Another project associated with *Perseus*, an electronic edition of the works of Christopher Marlowe, uses hyperlinks in an entirely different manner (Crane, 1998). It provides the full text of twenty different editions of Marlowe, ranging from 1590 to 1973 (see

FIGURE 5.7. *Perseus Project* (Sample Latin Text)

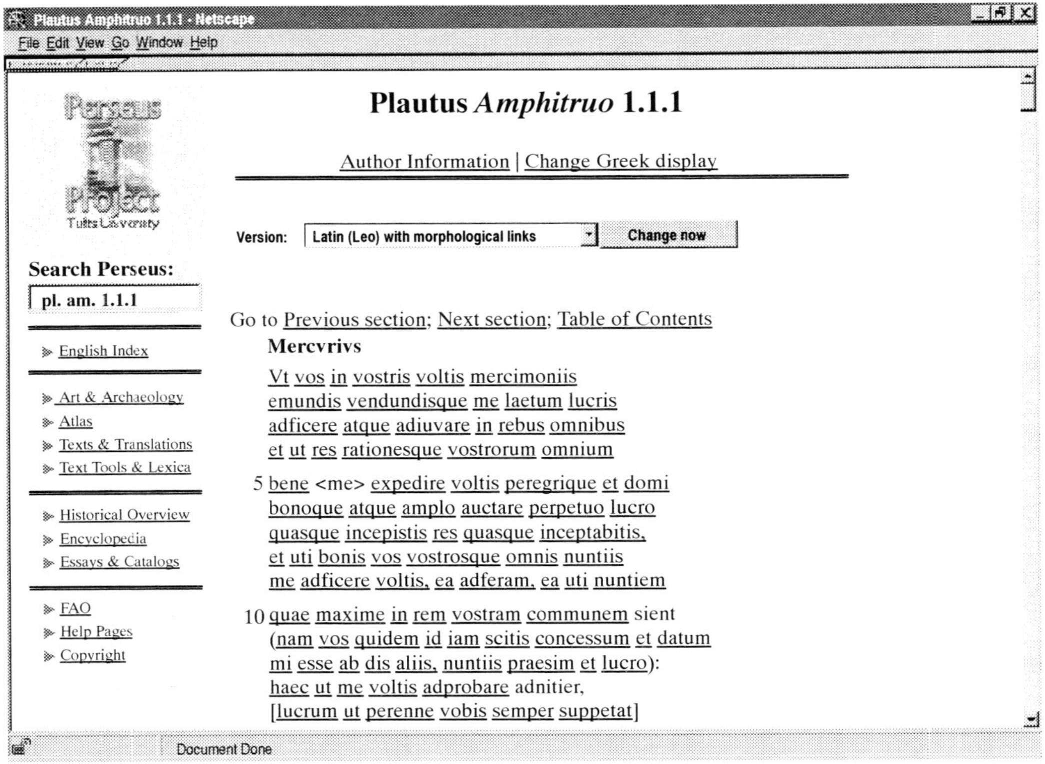

File Edit View Go Window Help

Perseus
project
Tufts University

Search Perseus:

pl. am. 1.1.1

» English Index

» Art & Archaeology
» Atlas
» Texts & Translations
» Text Tools & Lexica

» Historical Overview
» Encyclopedia
» Essays & Catalogs

» FAQ
» Help Pages
» Copyright

Document Done

Plautus *Amphitruo* 1.1.1

Author Information | Change Greek display

Version: | Latin (Leo) with morphological links | Change now

Go to Previous section; Next section; Table of Contents

Mercvrivs

Vt vos in vostris voltis mercimoniis
emundis vendundisque me laetum lucris
adficere atque adiuvare in rebus omnibus
et ut res rationesque vostrorum omnium

5 bene <me> expedire voltis peregrique et domi
bonoque atque amplo auctare perpetuo lucro
quasque incepistis res quasque inceptabitis,
et uti bonis vos vostrosque omnis nuntiis
me adficere voltis, ea adferam, ea uti nuntiem

10 quae maxime in rem vostram communem sient
(nam vos quidem id iam scitis concessum et datum
mi esse ab dis aliis, nuntiis praesim et lucro):
haec ut me voltis adprobare adnitier,
[lucrum ut perenne vobis semper suppetat]

143

Figure 5.8). These are interlinked in such a way that all the variants can be seen at any place in the text. Within each edition of a play, each passage is accompanied by marginal notes which list all the other editions with variants on that passage. These lists are hyperlinks to the variant text of the passage. There is also a mechanism for swapping between editions of each scene, using a pull-down menu that lists all twenty versions.

The *William Blake Archive*, which is being produced by the Institute for Advanced Technology in the Humanities at the University of Virginia, uses a complex hypertextual structure based mainly on images of the pages of Blake's illuminated books. These can be browsed within a hierarchical tree mirroring the page sequence of the physical originals. But other representations of each page can be summoned, such as enlargements of the page image and transcriptions of the page's text. These appear, not as sequential files on the screen, but as parallel windows that can be viewed alongside the original image. A search can also be invoked, either of the words in the text or of the contents of an image, using extensive descriptions and annotations that accompany each image. The images themselves can also be enlarged or reduced.

The Searchable Database

Although the weblike hypertext is undoubtedly the most typical and distinctive form of electronic text, there is one more type of structure to be considered. In this approach, the browsable text is dispensed with altogether. Instead, there is only a collection of textual resources that can be searched for particular combinations of letters, words, or characters. Technically speaking, this is a database, since it consists of a collection of data organized into structured fields for the purpose of random searching. Although databases are usually envisaged more in terms of bibliographical, biographical, or statistical information, electronic full-text files blur such distinctions. The text is fragmented into a collection of data that can be searched similar to any other database.

The *Dartmouth Dante*, directed by Robert Hollander, exemplifies this approach (see Figure 5.9). It contains the text of Dante's *Divine Comedy* and of forty-six commentaries on it, dating from 1322 to 1982. However, it is not possible to call up the entire poem or an

FIGURE 5.8. *Perseus Project*: Christopher Marlowe, *Doctor Faustus* (Sample)

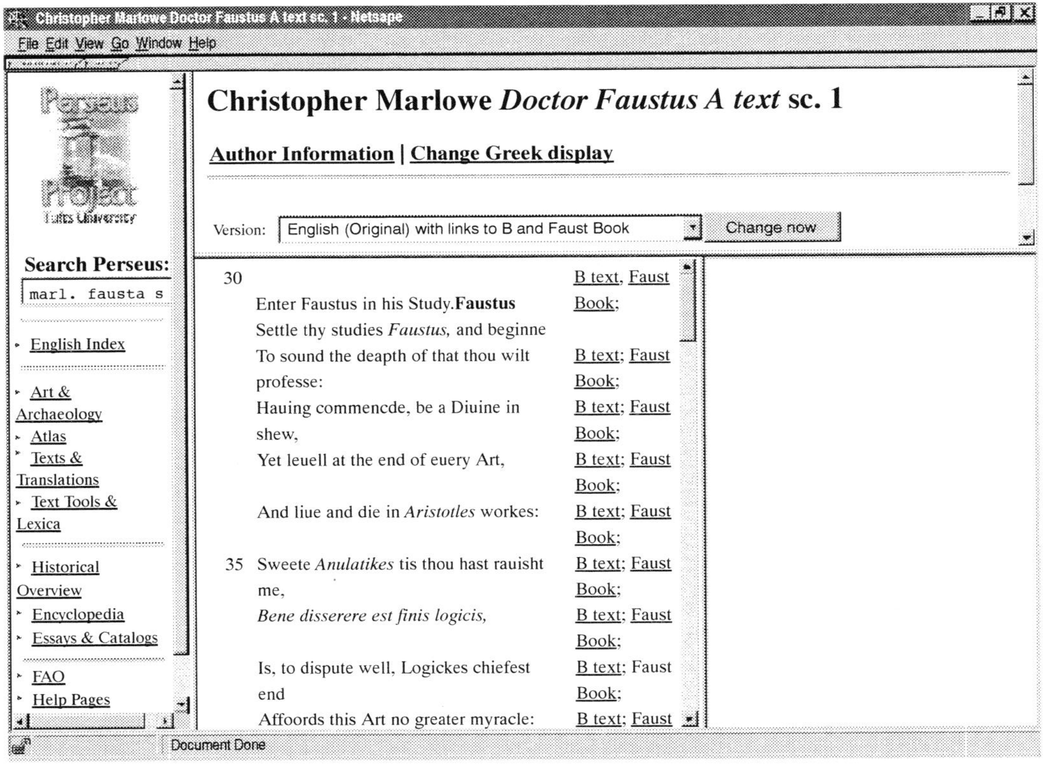

Christopher Marlowe Doctor Faustus A text sc. 1 - Netscape

File Edit View Go Window Help

Perseus Project
Tufts University

Search Perseus:

`marl. fausta s`

- English Index

- Art & Archaeology
- Atlas
- Texts & Translations
- Text Tools & Lexica

- Historical Overview
- Encyclopedia
- Essays & Catalogs

- FAQ
- Help Pages

Document Done

Christopher Marlowe *Doctor Faustus A text sc. 1*

Author Information | **Change Greek display**

Version: English (Original) with links to B and Faust Book   Change now

30   Enter Faustus in his Study,**Faustus**

Settle thy studies *Faustus*, and beginne

To sound the deapth of that thou wilt professe:

Hauing commencde, be a Diuine in shew,

Yet leuell at the end of euery Art,

And liue and die in *Aristotles* workes:

35   Sweete *Anulatikes* tis thou hast rauisht me,

Bene disserere est finis logicis,

Is, to dispute well, Logickes chiefest end

Affoords this Art no greater myracle:

B text: Faust Book;

B text: Faust Book;

B text: Faust Book;

B text: Faust Book;

B text: Faust Book;

B text: Faust Book;

B text: Faust Book;

B text: Faust Book;

B text: Faust

145

FIGURE 5.9. *Dartmouth Dante* (Search Screen)

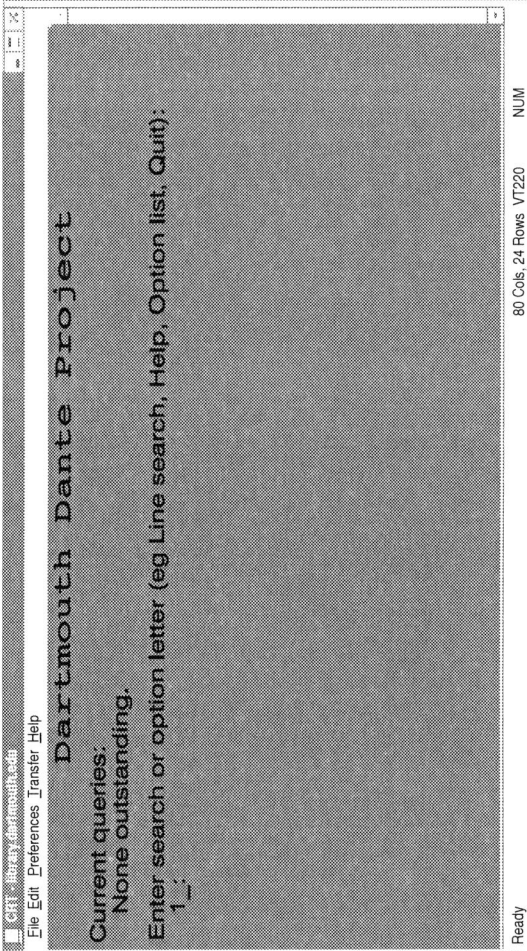

```
CRT - library.dartmouth.edu
File  Edit  Preferences  Transfer  Help

               Dartmouth Dante Project

Current queries:
  None outstanding.

Enter search or option letter (eg Line search, Help, Option list, Quit):
  1_:



Ready                                    80 Cols, 24 Rows  VT220          NUM
```

entire commentary on the screen and to scroll through it. The only entry point is a search screen, and the only way of reading the text is to retrieve a small fragment containing one or more search terms.

The *Dartmouth Dante* uses a comparatively sophisticated search language known as BRS/Search, which enables various types of searches:

- Searching the entire database for a specific word
- Searching for phrases
- Searching for word stems and pattern matching
- Combining searches
- Limiting searches to specific parts of the text, such as a particular cantica, canto, or even line reference
- Limiting searches to commentaries in a particular language

The result is a powerful analytical tool that can be used to ask complicated questions about the contents of a text. Software such as DynaText and Open Text also take this approach and allow most of the same searches to be made across large texts and corpora of texts.

This approach is rarely presented as the only approach to a text, however, and is usually found in parallel with another kind of structure, especially a hypertextual one. The text as a database, and nothing more than a database, violates our sense of reading. Most of us need to be able to browse a text in some way, instead of merely searching it in a manner that lacks all the contextual clues to be found in a readable text. Even the most fragmented and discontinuous hypertext retains enough of the sense of the wholeness of a work to satisfy this need. If the distinction between a text and a database still has any meaning at all, the *Dartmouth Dante* is not really a text. It assumes the existence of a well-known linear text, Dante's *Divine Comedy*, and chooses to provide only an index or concordance to it in electronic form. It effectively deconstructs the text into a collection of linguistic units.

Multiple Structures

Ultimately, there are only two fundamental architectures for electronic texts: the browsable text and the searchable text. Browsing can encompass simple linear scrolling, parallel browsing of multiple texts and versions of the same text, browsing from a table of contents, and browsing through a hypertext web. Searching can cover the text as a searchable database and the text as a component within a searchable corpus of material.

The most sophisticated electronic texts contain a variety of different structures simultaneously. A good example is Peter Robinson's CD-ROM edition of the *Prologue* to *The Wife of Bath's Tale,* produced as part of the *Canterbury Tales Project* (see Figure 5.10). It contains a table of contents, a text segmented into browsable sections, searchable databases, and hypertext links at numerous points in the text, especially to variant readings and transcriptions. There are also images linked to the text that can be browsed separately. Among its distinctive features are dozens of multiple versions of the text presented in parallel, as well as spelling databases covering all morphological forms. The text can be read in a linear order, browsed by myriad different routes, or searched for occurrences of specific letters or words. It is simultaneously a series of parallel linear texts, a hypertext web of interlocking text fragments, and a database text.

TYPES OF EDITIONS

Scholarly editions have been the subject of almost innumerable debates and discussions about approaches and methodologies, going back into the nineteenth century. There are widely divergent views about how to construct a critical edition of a work; these tend to crystallize around the extent to which the actual surviving texts of the work can be emended and what authority can be used as the source for choosing among variant readings. An important point of divergence is whether an edition should be eclectic or not. An eclectic text is one that combines readings from different sources— manuscripts, published editions, or the editor's conjectures—to produce a text that may never have previously existed as a historical reality. It is an editor's construction of an ideal text.

Peter Shillingsburg (1996:16-24) identifies five basic orientations underlying all the debates and discussions about scholarly editing. He emphasizes that this is a formal, theoretical analysis, not

FIGURE 5.10. *Canterbury Tales Project: The Wife of Bath's Prologue* (Opening Screen)

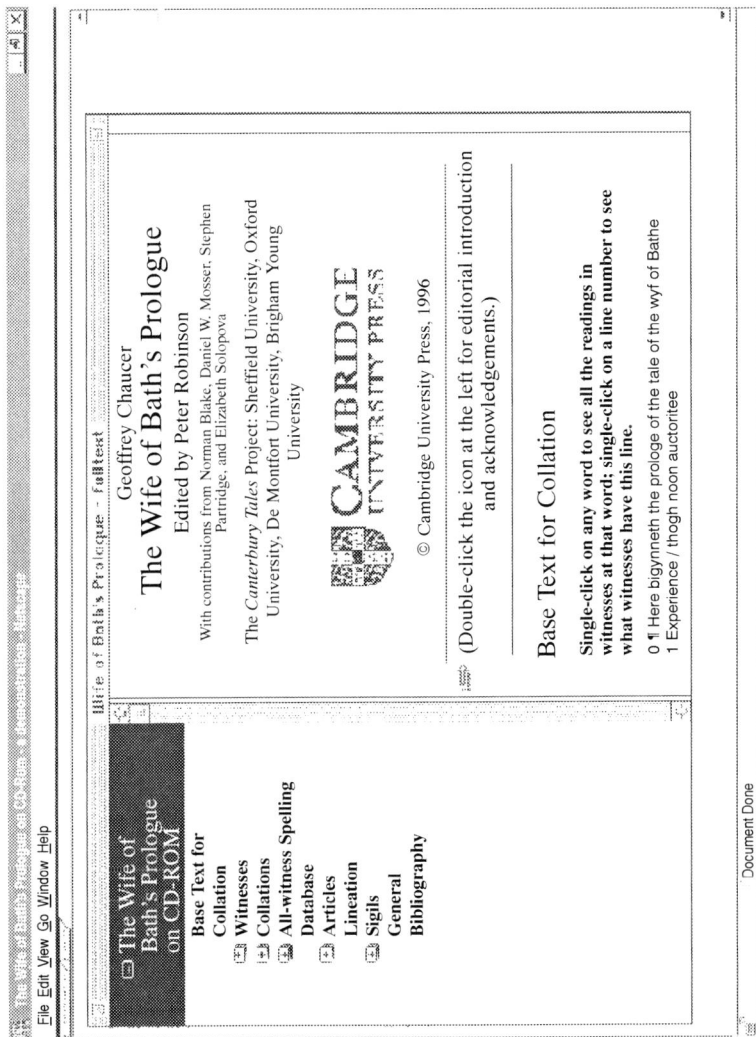

a classification of actual published types of editions. These five orientations are as follows:

- *Documentary* (or historical). Historically discrete texts must be kept separate rather than being merged into an eclectic critical edition. The aim is to preserve the "textual integrity of historical moments and physical forms" (Shillingsburg, 1996:17) and to present the documents in their historical form. The source of authority is the document itself.
- *Aesthetic.* This approach encourages the emendation of the text in the historical document by reference to the editor's or author's concept of artistic and stylistic correctness. The result is an eclectic text based on the authority of this concept of artistic forms.
- *Authorial.* The overriding concern of this orientation is to establish a text that reflects the intentions of the author. Authorial forms are preferred to nonauthorial forms, and the author's manuscript, if it exists, is preferred as a copy text. This approach has been very influential, with W.W. Greg, Fredson Bowers, and Thomas Tanselle as its main proponents, but has attracted considerable opposition and disagreement in recent years. It results in an eclectic text based on the authority of the editor's interpretation of the author's intentions.
- *Sociological.* Here, the social processes and institutions involved in publishing the work take precedence over the author's intentions. Editors who encourage this approach, such as Jerome McGann, tend to prefer a published text as their copy text and are unlikely to produce an eclectic text. The "social event that the text represents" and the "institutional unit" of the author and publisher are the preferred sources of authority (Shillingsburg, 1996:23).
- *Bibliographic.* This is more oriented toward analyzing texts and editions than toward providing a basis for editorial methodology. Associated particularly with D.F. McKenzie, the bibliographic approach investigates all the different aspects of the surviving physical forms of the text, including paper, typeface, binding, layout, and so on. Its insights can be applied particu-

larly in the sociological approach to editing and in relation to the analysis of scholarly editions themselves.

In practice, one or more of these different editorial orientations may overlap and coexist in any given scholarly edition. A different typology, based on actual published editions, is given by David Greetham (1994:347-355):

- *Photographic reprint:* a reproduction of a particular version of the text using photographs of the manuscript or print original
- *Type facsimile:* a reproduction of the physical appearance of a particular version of the text by resetting and reprinting using the same typeface as the original printing
- *Diplomatic transcript:* an exact transcription of a document using the same spelling, punctuation, and capitalization, but in a modern typesetting
- *Critical editions with inclusive text:* an eclectic text with the apparatus included on the same page as the text and keyed to it
- *Eclectic clear text with multiple apparatus:* an eclectic text, but with the apparatus entirely separate from the text and not keyed to it
- *Parallel text:* different versions of the text on facing pages
- *Genetic edition:* different layers in the genesis of a particular document integrated into a single critical text
- *Synoptic edition:* different layers in the genesis of a particular work integrated into a single critical text
- *Variorum edition:* a copy text reprinted, accompanied by an account of subsequent critics' and editors' emendations
- *Critical commentary:* a copy text reprinted, accompanied by the comments of the compiler and earlier critics

The first three and the last two of these are not "critical" editions, in that their compilers do not attempt to construct an authoritative text by comparing the surviving documents.

Debates in the last few years have tended to pit the documentary and sociological editors against their more traditional authorial colleagues. Critical and eclectic editions have been criticized for their presumption in claiming to know the author's intentions and for the alleged artificiality of the text they present. This criticism has some affinities with the postmodernist attack on what Roland Barthes (1986:52-53) called the "Author-God" and its rejection of the idea of a text as a closed, definitive utterance of the author/creator to a largely passive reader. Instead, the text is seen as a "collection of fragments" collected and arranged by the author, which can be read in multiple ways by the reader, who plays a much more active role in creating the text and may even be thought of as its cocreator (Tuman, 1992:64). When these ideas are applied to scholarly editing, the noneclectic approaches are preferred because of their emphasis on the actual documents containing the text and its social context.

Electronic editions are sometimes assumed to have a close connection with these postmodernist approaches to editing. Critics such as David Greetham have praised the ability of the electronic medium to provide an eclectic, open-ended, and "reader-driven" edition. "Most scholars now recognize," writes Greetham, "that there is a natural affinity between the computer and the variable discourses of contemporary textual scholarship" (Greetham, 1994:345). For Jay David Bolter (1991:3-6), the electronic edition changes the relationship between the author and the text, and between both of them and the reader. It "tends to reduce the distance between author and reader by turning the reader into an author" (Bolter, 1991:3).

We may well agree that the electronic medium is particularly suited to these kinds of ideas and approaches. It does not necessarily follow, however, that an electronic text must take a postmodernist form. Electronic texts can in fact take any of a variety of editorial approaches, and most of the possibilities are exhibited in the texts presently available. The most common approach is the single-copy text, but other types of editions are increasingly being explored, particularly facsimiles and documentary transcriptions.

Single-Copy Text

Many electronic texts simply recycle the text of an earlier printed edition. The archives of ASCII texts, such as *Project Gutenberg*, tend to rely on copy texts derived from a printed version chosen only for its ready availability and lack of copyright ownership. In many cases, the source of the copy text is not documented. By its own account, *Project Gutenberg* "has avoided requests, demands, and pressures to create authoritative editions. We do not write for the reader who cares whether a certain phrase in Shakespeare has a ':' or a ';' between its clauses. We put our sights on a goal to release etexts that are 99.9% accurate in the eyes of the general reader" (Project Gutenberg).

The attitude of some in the scholarly editing community toward these texts is well summarized by Peter Shillingsburg (1996:161). He says that *Project Gutenberg* is "the product of abysmal ignorance of the textual condition" and emphasizes the unreliability of its texts, which are insufficiently proofread, lacking in documentation of their sources, and missing information on formatting and pagination. But other scholars are less critical, believing that *Project Gutenberg* at least makes texts available to people and countries who do not have easy access to printed versions, and provides a basic electronic form of important texts that can be improved and used for other purposes.

Commercial publishers of electronic texts generally also use single-copy texts taken from printed editions that are now out of copyright. Chadwyck-Healey Ltd. follows this approach consistently. For medieval and patristic works, its source is the vast nineteenth-century collection assembled by the abbé Migne in his *Patrologia Latina*. In many cases, these simply reproduce even earlier editions and are very variable in quality. Although Migne's collection is out of copyright, many of the texts have been subsequently reedited and significantly amended. In some cases, the authorship of a text has been reattributed to a different writer. The only concession to this in the electronic version is the addition of a code to indicate where the authorial attribution has been changed in later printed editions.

A similar procedure is adopted for Chadwyck-Healey's various databases of English literary works. These include *English Poetry*, which reproduces 4,500 printed sources, and *English Drama*, which contains more than 4,000 different plays. The printed editions used as the basis for these electronic versions are, for the most part, pre-twentieth century, out of copyright, and contemporary with the author. In all cases, however, the copy text is carefully documented, and its physical features are recorded in the electronic version.

More recently, Chadwyck-Healey has begun to use twentieth-century editions as copy texts for its electronic versions of modern poetry. This is the result of negotiations with the publishers and copyright holders of these works. The American company Intelex has followed a similar approach with its series of philosophy texts, distributed under the title *Past Masters*. These generally use as their copy text a twentieth-century critical edition, under agreement with the publishers of the printed versions. The *Past Masters* version of the published works of Ludwig Wittgenstein, for example, reproduces in electronic form the definitive editions of his various writings in German and English, issued by Oxford University Press between 1961 and 1980, as well as various journal articles by him.

Collections of noncommercial electronic texts, such as those at the University of Virginia's Electronic Text Center or the Oxford Text Archive, also tend to use a single printed version of the text as their base. This may be either a critical edition or just a convenient printed text, but the source is usually carefully documented. Information about the physical features of the original is normally included with the electronic text. In the case of the Oxford Text Archive, which contains texts from a variety of different sources, this standard has not always been observed. The University of Virginia's extensive collections of electronic texts are much more standardized, having been created in-house in accordance with a fixed set of rules and conventions. These include the following:

- With only a few exceptions the print source of the text is clearly noted. Generally, Virginia does not provide access to a text whose provenance is unknown.
- Incoming electronic texts are usually spell-checked, or verified against the original text.
- Details of typography are generally maintained—special characters and italics in particular can have a real bearing on meaning and should not be dropped from the electronic version of a text.
- Chapter divisions and pagination are retained for purposes of reference. Paginating a text is also a good way to check that it is complete.

• Illustrations are normally included, as they are an important part of the experience of reading the work.

The *Victorian Women Writers Project* also follows this general approach. Its goal is to produce accurate transcriptions of literary works by British women writers of the nineteenth century, including anthologies, novels, political pamphlets, and volumes of poetry and verse drama. An important consideration for the project is the accuracy and completeness of the texts. It also aims to ensure that an accurate bibliographical description of each source text is included, together with notes on the methods of transcription used. The intention is to provide an electronic representation of the original nineteenth-century editions.

Facsimile Editions

Facsimiles are an important type of printed edition in which an original text is reproduced photographically. This is usually applied to manuscripts and can reach extraordinarily lavish standards. The recent facsimile of the Ellesmere manuscript of Chaucer is a marvelous example of this craft, which is remarkably faithful to the original in size, colors, and pagination. At the other extreme, perhaps, are the black and white reproductions found in numerous microfilm and microfiche sets, which are adequate for basic study but not for detailed scholarly research and analysis. In the electronic context, facsimiles consist of a series of digital images at various levels of resolution and with varying degrees of color reproduction.

The *Electronic Beowulf Project*, directed by Kevin Kiernan, is a good example of a facsimile edition in electronic form. It contains hundreds of color images of the single surviving manuscript of the *Beowulf* poem, which dates from the early eleventh century. The fiberoptic and ultraviolet techniques used for these images are of such sophistication that they have revealed letters which were made unreadable by fire damage to the manuscript in 1731. The project also contains images of important eighteenth-century transcriptions of the manuscript, made before the fire, as well as images of selections from important nineteenth-century editions, translations, and collations. These images are far more numerous and accurate than would ever be possible with a printed facsimile edition. The underlying purpose of the electronic version, however, is generally the same as a printed facsimile: to reproduce the original manuscript or text in a way that is as faithful to the original as possible.

There are some similarities in the work of the *William Blake Archive*. Its main content is a database of about 3,000 images of individual pages in Blake's nineteen illuminated books. Each book is represented by several copies, giving a total of about fifty-five of the 195 hand-colored copies known to have been produced by Blake himself. These are accompanied by images taken from his paintings, drawings, and engravings. The value of such facsimiles is enhanced by transcriptions of the text and descriptions of the contents of each image by the editors. These can be searched for particular words, themes, and subjects. A further enhancement is the use of specialized software for enlarging, reducing, and manipulating the images. All these features go well beyond what is possible in a printed facsimile edition.

Documentary Transcriptions

Electronic texts are particularly suited to the provision of transcriptions of multiple documents that provide evidence for the different states of a particular work. In a printed edition, it is usually prohibitively expensive to provide full parallel transcriptions of multiple documents. For many literary works, this would involve one or more authorial manuscripts, as well as several printed editions and possibly proof copies as well. For a widely circulated medieval work, there are likely to be dozens of documents or witnesses to the text, each with its own set of variants. In a printed edition, this variety is usually reduced to an apparatus linking variants to a base text or to a transcript or facsimile of a couple of the most important manuscripts. But in the electronic context, the cost of presenting full transcriptions of each of the witnesses in parallel is far less, at least as far as the computer storage is concerned. The only significant cost arises from the time required to transcribe each witness.

Hence, most major projects that aim at electronic editions of medieval works are planning to provide full, parallel transcriptions of as many of the surviving witnesses as possible. Peter Robinson's *Canterbury Tales Project* is one of the most ambitious on this count.

It is transcribing all the witnesses from before 1500 to the text of *The Canterbury Tales*, both the surviving manuscripts and the earliest printed editions. There are eighty-eight of these, of which about sixty are relatively complete. Although the computer storage required is far from trivial, its cost is only a fraction of what a printed equivalent would require.

A similar process, for a later period, can be seen in the database of *Editions and Adaptations of Shakespeare* published by Chadwyck-Healey Ltd. This contains transcriptions of eleven major editions of the works of Shakespeare published from the First Folio of 1623 through to the Cambridge edition of 1863 to 1866. Also included are transcriptions of twenty-four original printings of individual plays, and more than a hundred adaptations, sequels, and burlesques. All these different versions can be easily compared and analyzed.

Multiple Approaches

In the most ambitious of the electronic text projects, the editors prefer to take more than one approach. Several types of editions are presented in parallel, usually interlinked in such a way that users can switch between them at will. These editions aim to be more than just "an undigested chaos" of documents (Shillingsburg, 1996:165). Instead they form a network of materials that are carefully related and linked to one another. For those editors who prefer a documentary or sociological approach to their task, this network can be left fairly open-ended. For editors with an authorial orientation, the network can include a critical, eclectic text as part of its resources.

Hoyt Duggan's electronic archive of *Piers Plowman* combines several types of editions. It contains documentary transcriptions of the eight main manuscript witnesses to the "B" version of Langland's fourteenth-century poem (out of seventeen surviving manuscripts of this version). These are accompanied by color digital facsimiles of these manuscripts and descriptions of them. There is also an annotated critical text based on one of the manuscripts (Cambridge, Trinity College B.15.17), as well as an attempt to reconstruct the archetype of the "B" version. Line-by-line collations of the manuscripts are available. The archive, in its final published form, will also be searchable.

The electronic edition of the *Prologue* to *The Wife of Bath's Tale*, prepared by Peter Robinson and his collaborators as the first stage in the *Canterbury Tales Project*, is another ambitious edition that combines several different approaches. It includes full documentary transcriptions of all fifty-eight witnesses to this portion of the text that date earlier than 1500. Each of these transcriptions is also accompanied by digital facsimiles, for a total of more than 1,100 pages of text. Also included are full collations of all the witnesses, in both regularized and unregularized form, as well as databases of variants and spellings. The copy text is the earliest witness—the Hengwrt manuscript, lightly amended—and this forms the base to which all the transcriptions are linked. There is no critical edition included in this electronic version, the editors preferring to present a definitive collection of materials to assist further critical work by scholars in the future.

Electronic Texts and Editorial Theory

The *Guidelines for Electronic Scholarly Editions* adopted by the Modern Language Association of America in 1997 acknowledge that different methods of editing may be appropriate in different situations. Rather than prescribing a particular method, the MLA Committee on Scholarly Editions encourages editors "to choose editorial procedures appropriate to their materials" and to experiment with a variety of approaches to electronic editions (MLA, 1997). The committee emphasizes, however, that an electronic edition should meet the same standards of accuracy, thoroughness, and consistency as a printed one. It does allow for the possibility that the usual editorial apparatus found in a printed edition may be replaced by full-text transcriptions, as long as there are mechanisms for displaying parallel texts and for collating textual variants.

The electronic text can reflect any of the major approaches to editing texts and offers the editor far more flexibility in presenting an edition than is possible within the confines of the printed page. It also seems to be true that open-ended, postmodernist editions need to use the electronic form to achieve the flexibility they require. Variant texts can be stacked for separate and simultaneous viewing. The multiple intentions of the author, and of subsequent readers, can be presented in parallel. The reader can construct a personal critical edition from the range of materials presented. In David Greetham's apt simile, these "fragmented, spliced, versioned, polysemous texts" are the equivalent of the Pompidou Centre in Paris, with all its structural elements exposed, or like a giant Meccano set (Greetham, 1993:14-15).

The issues are particularly acute when related to medieval texts. This is because variants are so endemic in the different witnesses to these texts. In Bernard Cerquiglini's famous dictum "medieval writing does not produce variants; it is variance" (1989:111). In an essay on *Piers Plowman* and *The Canterbury Tales*, Derek Pearsall (1985:97) speaks of the "tyranny of the critical edition," which imposes a single, but mythical, text on a complex reality of widely varying documentary witnesses. He suggests that "facsimiles, or at least seriatim transcripts, of single manuscripts" would be more authentic than a "sterilised" critical edition (Pearsall, 1985:105).

As long ago as 1989, Cerquiglini speculated that the computer might offer an effective way of showing the inherent variance of medieval texts. It can combine a multidimensional approach, using multiple windows for multiple witnesses, and an interactive presentation, enabling the reader to engage in a dialogue with the text. He sees the same characteristics of variance and mobility in the electronic medium and claims that an electronic text can rediscover "the path of an older literature, whose traces have been erased by the printed book" (Cerquiglini, 1989:116). Paradoxically, then, the typical instrument of the era of the "après-texte" can show us an image of the variance inherent in the era of the "pre-text."

In arguments such as Cerquiglini's, the electronic text becomes something of a peg on which to hang a critique of the scholarly edition in its printed embodiment. It may well be the case that an electronic edition is necessary to break the shackles of the traditional critical edition and to usher in new approaches to scholarly editing. But the electronic text can also subsume and incorporate most, if not all, of the many different forms of its printed predecessors. Although it will never replace the book as a vehicle for a convenient reading text, it seems increasingly likely to become the preferred medium for presenting the results of extensive scholarly study of a particular work.

FURTHER READING

Essential reading on the relationship between electronic texts and printed ones includes Bolter (1991) and Tuman (1992). Lanham (1993) is a passionate advocate of the electronic text. Nunberg (1996) contains a range of articles on the future of the book.

Williamson (1983) provides an excellent account of the design of printed books, while Steinberg (1974) is the standard history of printing. Turner (1977) is the definitive study of the emergence of the codex.

The best recent guide to textual scholarship is Greetham (1994). Other essential works on scholarly editing include Shillingsburg (1996), Bornstein and Williams (1993)—especially the chapters by Gabler, Greetham, and Shillingsburg—and several of the chapters in Eggert (1990).

On editing medieval texts specifically, there are numerous books and articles, among which Cerquiglini (1989) is important, and Masters (1992) is also interesting and relevant. Pearsall's 1985 article is particularly stimulating. Tanselle (1983) offers a rather different point of view.

Much of what is written about "hypertext" and "hypermedia" is hype. Worthy of serious attention are Delaney and Landow (1991), Conner (1992), and McGann (1995).

Useful short overviews of electronic editions are given by Barwell (1995), Robinson (1993b), and Faulhaber (1991). The papers in Finneran (1996) are also important.

WEB SITES

Academic Press—*IDEAL:* http://www.idealibrary.com/

Australian Cooperative Digitisation Project:
http://www.nla.gov.au/acdp/serials.html

Canterbury Tales Project:
http://www.shef.ac.uk/uni/projects/ctp/index.html

Chadwyck-Healey Ltd: http://www.chadwyck.co.uk/
Literature Online: http://lion.chadwyck.co.uk/
Patrologia Latina: http://pld.chadwyck.co.uk/

Dartmouth Dante:
http://www.nyu.edu/library/bobst/research/etc/dante.htm

The Electronic Beowulf:
http://www.uky.edu/~kiernan/BL/kportico.html

Electronic Binding Project (Ebind):
http://sunsite.berkeley.edu/Ebind/

Intelex—*Past Masters*: http://www.nlx.com/pstm/index.htm

Internet Public Library: http://www.ipl.org/

James J. O'Donnell: http://ccat.sas.upenn.edu/jod/

Labyrinth Library of Medieval Studies:
http://www.georgetown.edu/labyrinth/labyrinth-home.html

Making of America: http://moa.umdl.umich.edu/moa/index.html

Oxford Text Archive: http://ota.ahds.ac.uk/

Perseus Project: http://www.perseus.tufts.edu/

Piers Plowman Project:
http://jefferson.village.virginia.edu/piers/tcontents.html

Project Gutenberg: http://promo.net/pg/

Project Muse (Johns Hopkins University Press):
http://muse.jhu.edu/muse.html

Rabanus Maurus (Schipper): http://www.mun.ca/rabanus/

University of Virginia Electronic Text Center:
http://etext.lib.virginia.edu/

Victorian Women Writers Project:
http://www.indiana.edu/~letrs/vwwp/

The William Blake Archive:
http://jefferson.village.virginia.edu/blake/main.html

Conclusion

At first glance, electronic texts may seem to require a daunting amount of technical knowledge. Different kinds of markup, various processes of keying and scanning, an array of specialized software—all must be understood and applied in order to create an effective electronic text. Methods for distributing the text, ranging from the simple diskette to the World Wide Web, need to be considered and compared. Issues relating to the preservation and cataloging of such texts are also important. Broader institutional and organizational settings for creating, collecting, and accessing electronic texts may have to be investigated and assessed. Within the text itself, there are questions of structure and design to be weighed and resolved.

To some extent, the apparent complexity of this picture arises from its sheer novelty and unfamiliarity. After all, printed texts also require a complex infrastructure of publishers, printers, and libraries. They raise issues of cataloging and preservation and exist in many different structures and designs. The familiarity and longevity of these solutions and approaches tend to disguise their underlying complications. They are also largely hidden from the creators and editors of texts because of the separate industry that exists to carry them out.

Even so, electronic texts undoubtedly involve a greater level of technical complexity than printed ones do. In the face of this, it is important to consider whether the extra effort is justified. What are the benefits of creating electronic texts? There are, in fact, a range of answers to this question. Most obviously, an electronic medium such as the World Wide Web can be used to make texts far more widely available. This is especially so for titles which are long out of print and which are available in printed form only in the largest and oldest libraries. Texts that exist in manuscript form will benefit greatly from publication in electronic form; far more people will have access to them, and there will be less pressure to use the often-fragile originals. In regions and countries that lack an adequate infrastructure of libraries and book distribution, even the more common texts may not be available in print. Electronic versions, in contrast, can make use of a growing infrastructure of computer networks that is being developed mainly for commercial purposes.

Even where there is an abundance of printed texts, electronic editions offer various unique and significant benefits. For a start, they make it possible to search a text, or a corpus of texts, by keyword and to analyze its contents in a variety of ways. The more detailed the markup and encoding, the more sophisticated this analysis can be. Searching for quotations, for uses of particular terms and phrases, for patterns within texts, and for connections between texts is far easier than with printed volumes. The *Patrologia Latina Database* is a case in point. The 221 volumes of the printed original can be searched in minutes—instead of months or even years.

Electronic editions also make it much more feasible to compare multiple versions of the same text. Publishing facsimiles or transcriptions in printed form is all but impossible, despite their interest and importance to scholars. In the electronic context, they are not only a reasonable proposition, but they can be compared and contrasted with much greater subtlety and sophistication. The *Canterbury Tales Project* and the *William Blake Archive*, among many others, clearly demonstrate the value of this approach.

An electronic text can also serve as the center of a collection of related materials in a variety of different media. Notes, commentaries, reference works, images, and audiovisual files can also be linked to the text in various ways. A well-designed digital library such as the *Perseus Project* illustrates the rich possibilities that are available here. Each individual text exists within a network of related materials that can be explored by numerous different paths.

Electronic texts are never likely to replace printed ones. But it is possible to do things in the electronic medium that are impossible, or uneconomic, to do in print: index every word in the text, transcribe every variant, and link the text to a videotaped performance, for instance. Above all, however, electronic versions can bring these texts to a new readership. We live in a visual age, where so much communication takes place through the computer and the television screen, where the image is more powerful than the text. By developing electronic texts that exist in a visual medium and are embedded in a visual context, we can make use of the power of the image to convey the importance of the text. From this point of view, electronic texts are a vitally important tool for ensuring that our great heritage of printed books and manuscripts retains its place in an increasingly visual and image-centered world.

Bibliography

Adobe Systems Inc. 1985. *PostScript language reference manual*. Reading, MA: Addison-Wesley Longman.

Alschuler, Liora. 1995. *ABCD . . . SGML: A user's guide to structured information*. London: International Thomson Computer Press.

Barthes, Roland. 1986. *The riddle of language*. New York: Hill and Wang.

Barwell, Graham. 1995. "Electronic editions: An overview," *Bibliographical Society of Australian and New Zealand Bulletin* 19(2):79-87.

Beagrie, Neil and Daniel Greenstein. 1998. "Digital collections: A strategic policy framework for creating and preserving digital resources": <http://ahds.ac.uk/manage/framework.htm>.

Besser, Howard and Jennifer Trant. 1995. *Introduction to imaging: Issues in constructing an image database*. Santa Monica, CA: Getty Art History Information Program.

Bolter, Jay David. 1991. *Writing space: The computer, hypertext, and the history of writing*. Hillsdale, NJ: Lawrence Erlbaum.

Borges, Jorge Luis. 1965. *Fictions*. London: John Calder.

Bornstein, George and Ralph G. Williams (Eds.). 1993. *Palimpsest: Editorial theory in the humanities*. Ann Arbor, MI: University of Michigan Press.

Bradley, Neil. 1997. *The concise <SGML> companion*. Harlow, Essex, UK: Addison-Wesley Longman.

Bryan, Martin. 1997. *SGML and HTML explained*, Second edition. Harlow, Essex, U.K.: Addison-Wesley Longman.

Buckland, Michael. 1994. "From catalog to selecting aid," *ALCTS Newsletter* 5(5):A-D.

Burnard, Lou and C.M. Sperberg-McQueen. 1995. "TEI Lite: An introduction to text encoding for interchange": <http://www.uic.edu/orgs/tei/intros/teiu5.html>.

Burrows, Toby. 1996. "Using DynaWeb to deliver large full-text databases in the humanities," *Computers & Texts* 13:15-17.

Burrows, Toby. 1997. "Textus ex machina: Electronic texts and medieval studies," *Parergon* 14(2):67-83.

Burrows, Toby. 1998. "Beyond HTML: Markup languages and the future of electronic information," *Australian Academic and Research Libraries* 29(2): 150-156.

Cerquiglini, Bernard. 1989. *Éloge de la variante: Histoire critique de la philologie*. Paris: Seuil.

Chaucer, Geoffrey. 1996. *The Wife of Bath's Prologue on CD-ROM*, Peter Robinson (Ed.). Cambridge: Cambridge University Press.

Chernaik, Warren, Caroline Davis, and Marilyn Deegan (Eds.). 1993. *The politics of the electronic text*. Oxford: Office for Humanities Communication, Publication No. 3, Oxford University Computing Services.

Conner, Patrick W. 1992. "Hypertext in the last days of the book," *Bulletin of the John Rylands Library* 74(3):7-24.

Connolly, Dan (Ed.). 1997. *XML: Principles, tools, and techniques*. Sebastopol, CA: O'Reilly. Also published as: *World Wide Web Journal* 2(4).

Coombs, James H., Allen H. Renear, Steven J. DeRose. 1987. "Markup systems and the future of scholarly text processing," *Communications of the ACM* 30(11):933-947.

Council on Library and Information Resources. 1998. "On the preservation of knowledge in the electronic age": <http://www.clir.org/film/discussion.html>.

Crane, Gregory. 1998. "The Perseus Project and beyond: How building a digital library challenges the humanities and technology." *D-Lib Magazine,* January. <http://www.dlib.org./january98/01crane.html>.

Davis, Stephen Paul. 1996. "SGML-MARC: Incorporating library cataloging into the TEI environment": <http://www.columbia.edu/cu/libraries/inside/projects/sgml/sgmlmarc/davis.9603.text.html>.

Delaney, Paul and George P. Landow (Eds.). 1991. *Hypermedia and literary studies*. Cambridge, MA: The MIT Press.

Delaney, Paul and John K. Gilbert. 1991. "HyperCard stacks for Fielding's *Joseph Andrews*: Issues of design and content." In Paul Delaney and George Landow (Eds.), *Hypermedia and Literary Studies*. Cambridge, MA: The MIT Press, 287-298.

DeRose, Steven J. 1993. "Markup systems in the present." In George Landow and Paul Delaney (Eds.), *The digital word*. Cambridge, MA: The MIT Press, 119-135.

Dougherty, Dale. 1997. "The XML files: Multidimensional files that go beyond HTML," *Web Review,* May 16: <http://webreview.com/97/05/16/feature/index.html>.

Dowler, Lawrence (Ed.). 1997. *Gateways to knowledge: The role of academic libraries in teaching, learning, and research*. Cambridge, MA: The MIT Press.

DuCharme, Bob. 1998. *SGML CD*. Upper Saddle River, NJ: Prentice-Hall PTR.

Eggert, Paul (Ed.). 1990. *Editing in Australia*. Canberra, Australia: English Department, University College ADFA.

Ellis, Steven. 1996. "Toward the humanities digital library: Building the local organization," *College & Research Libraries* 57(6):525-534.

Ensign, Chet. 1997. *$GML: The billion dollar secret*. Upper Saddle River, NJ: Prentice-Hall PTR.

Faulhaber, Charles. 1991. "Textual criticism in the 21st century," *Romance Philology* 45(1):123-148.

Finneran, Richard J. (Ed.). 1996. *The literary text in the digital age*. Ann Arbor, MI: University of Michigan Press.

Fleischhauer, Carl. 1996. "Digital formats for content reproductions": <http://lcweb2.loc.gov/ammem/formats.html>.

Fleishman, Glenn. 1997. "A closer view of Web graphics": <http://www.netbits.net/nb-issues/NetBITS-007.html>.

Flynn, Peter. 1997a. "Commonly asked questions about the Extensible Markup Language: The XML FAQ": <http://www.ucc.ie/xml/>.

Flynn, Peter. 1997b. "W[h]ither the Web? The extension or replacement of HTML," *Journal of the American Society for Information Science* 48(7): 614-621.

Gaynor, Edward. 1994. "Cataloging electronic texts: The University of Virginia Library experience," *Library Resources & Technical Services* 38(4):403-413.

Gaynor, Edward. 1996. "From MARC to markup: SGML and online library systems," *ALCTS Newsletter* 7(2):A-D. Also at: <http://www.lib.virginia.edu/speccol/scdc/articles/alcts.brief.html>.

Giordano, Richard. 1995. "The TEI header and the documentation of electronic texts." In Nancy Ide and Jean Véronis (Eds.), *Text encoding initiative: Background and context.* Dordrecht, Netherlands: Kluwer, 75-84.

Goldfarb, Charles F. 1990. *The SGML handbook.* Oxford: Clarendon Press.

Goldfarb, Charles F., Steve Pepper, and Chet Ensign. 1997. *SGML buyer's guide: A unique guide to determining your requirements and choosing the right SGML and XML products and services.* Upper Saddle River, NJ: Prentice-Hall PTR.

Graham, David. 1991. "The emblematic hyperbook." In Paul Delaney and George Landow (Eds.), *Hypermedia and literary studies.* Cambridge, MA: The MIT Press, 273-286.

Graham, Ian S. 1996. *HTML sourcebook*, Second edition. New York: Wiley.

Graham, Peter. 1995. "Requirements for the digital research library," *College & Research Libraries* 56(4):331-339.

Greetham, D.C. 1993. "Editorial and critical theory: From modernism to postmodernism." In George Bornstein and Ralph Williams (Eds.), *Palimpsest: Editorial theory in the humanities.* Ann Arbor, MI: University of Michigan Press, 9-28.

Greetham, D.C. 1994. *Textual scholarship: An introduction.* New York: Garland.

Guenther, Rebecca S. 1996. "The challenges of electronic texts in the library: Bibliographic control and access." In Robin P. Peek and Gregory B. Newby (Eds.), *Scholarly publishing: The electronic frontier.* Cambridge, MA: The MIT Press, 251-275.

Hockey, Susan. 1996. "Creating and using electronic editions." In Richard Finneran (Ed.), *The literary text in the digital age.* Ann Arbor, MI: University of Michigan Press, 1-21.

Hume, David. 1990. *HUMETEXT 1.0*, prepared by T.L. Beauchamp, D.F. Norton, and M.A. Stewart. (Place and publisher unknown.)

Ide, Nancy M. and C.M. Sperberg-McQueen. 1995. "The TEI: History, goals, and future." In Nancy Ide and Jean Véronis (Eds.), *Text Encoding Initiative: Background and context.* Dordrecht, Netherlands: Klewer, 5-15.

Ide, Nancy M. and Jean Véronis (Eds.). 1995. *Text Encoding Initiative: Background and context*. Dordrecht: Kluwer. Also published as: *Computers and the humanities* 29(1/3).

Karpinski, Richard. 1997. "New browser options emerge," *Internet Week*, November 26: <http//www.internetwk.com/news/news1126-9.htm>.

Kenney, Anne R. and Stephen Chapman. 1996. *Digital imaging for libraries and archives*. Ithaca, NY: Department of Preservation and Conservation, Cornell University Library.

Lancashire, Ian, J. Bradley, W. McCarty, M. Stairs, T.R. Wooldridge. 1996. *Using TACT with electronic texts*. New York: Modern Language Association of America.

Landow, George P. and Paul Delaney (Eds.). 1993. *The digital word: Text-based computing in the humanities*. Cambridge, MA: MIT Press.

Lanham, Richard A. 1993. *The electronic word: Democracy, technology, and the arts*. Chicago: University of Chicago Press.

Lee, Stuart. 1996. "The Internet and the humanities scholar." In Christine Mullings, Marilyn Dugan, Seamus Ross, and Stephanie Kenna (Eds.), *New technology for the humanities*. London: Bowker-Saur, 426-441.

Leslie, Michael. 1993. "Electronic editions and the hierarchy of texts." In Warren Chernaik, Caroline Davis, and Marilyn Deegan (Eds.), *The politics of the electronic text*. Oxford: Office for Humanities Communication, Publication No. 3, Oxford University Computing Services, 41-51.

Library of Congress. 1995. "Mapping the Dublin Core metadata elements to USMARC" (USMARC Advisory Group discussion paper, No. 86): <gopher://marvel.loc.gov/00/.listarch/usmarc/dp86.doc>.

Lie, Hakon Wium and Bert Bos. 1997. *Cascading Style Sheets: Designing for the Web*. Harlow, Essex, U.K.: Addison-Wesley Longman.

Light, Richard. 1997. *Presenting XML*. Indianapolis, IN: Sams.net.

Lowry, Anita. 1997. "Gateways to the classroom." In Lawrence Dowler (Ed.), *Gateways to knowledge: The role of academic libraries in teaching, learning, and research*. Cambridge, MA: The MIT Press, 199-206.

Lyman, Peter. 1997. "The gateway library: Teaching and research in the global reference room." In Lawrence Dowler (Ed.), *Gateways to Knowledge: The role of academic libraries in teaching, learning, and research*. Cambridge, MA: The MIT Press, 135-147.

Lynch, Clifford. 1997a. "Searching the Internet," *Scientific American* March. <http://www.sciam.com/0397issue/0397lynch.html>.

Lynch, Clifford. 1997b. "The Z39.50 information retrieval standard. Part I: A strategic view of its past, present and future," *D-Lib Magazine* April.

Mace, Scott, Udo Flohr, Rick Dobson, and Tony Graham. 1998. "Weaving a better Web," *Byte* March: <http://byte.com/art/9803/sec5/art1.htm>.

Mackenzie, David. 1986. *A manual of manuscript transcription for the dictionary of the Old Spanish language,* Fourth edition. Madison, WI: Hispanic Board of Medieval Studies.

Masters, Bernadette A. 1992. *Esthétique et manuscripture: Le "moulin à paroles" au moyen âge.* Heidelberg, Germany: Carl Winter.

Matthews, Martin S. and Erik B. Poulsen. 1998. *FrontPage 98: The complete reference.* Berkeley, CA: Osborne McGraw-Hill.

McClelland, Deke and John San Filippo. 1997. *PageMill 2 for dummies.* Foster City, CA.: IDG Books.

McGann, Jerome. 1995. "The rationale of hypertext" <http://lists.village.virginia.edu/public/jjm2f/rationale.html>.

Miller, Paul and Daniel Greenstein (Eds.). 1997. "Discovering online resources across the humanities: A practical implementation of the Dublin Core": <http://ahds.ac.uk/public/metadata/discovery.html>.

Modern Language Association of America. 1997 "Guidelines for electronic scholarly editions": <http://sunsite.berkeley.edu/MLA/intro.html>.

Mullings, Christine, Marilyn Deegan, Seamus Ross, and Stephanie Kenna (Eds.). 1996. *New technologies for the humanities.* London: Bowker-Saur.

Nelson, Theodor Holm. 1987. *Literary machines.* South Bend, IN: Author.

Nordenfalk, Carl. 1977. *Celtic and Anglo-Saxon painting.* London: Chatto and Windus.

Nunberg, Geoffrey (Ed.). 1996. *The future of the book.* Berkeley, CA: University of California Press.

O'Gorman, Lawrence and Rangachar Kasturi. 1995. *Document image analysis.* Los Alamitos, CA: IEEE Computer Society Press.

Olsen, Jan. 1997. "The gateway: Point of entry to the electronic library." In Lawrence Dowler (Ed.), *Gateways to knowledge: The role of academic libraries in teaching, learning, and research.* Cambridge, MA: The MIT Press, 123-134.

Oxford Text Archive. 1997. "Annual report of the Oxford Text Archive 1996-97": <http://sable.ox.ac.uk/reports/annrep_main.html>.

Palmer, Pete, Adam Schneider, Anne Chenette. 1996. *The Web server handbook.* Upper Saddle River, NJ: Prentice-Hall PTR.

Pattie, Ling-Yuh W. and Bonnie Jean Cox (Eds.). 1996. *Electronic resources: Selection and bibliographic control.* Binghamton, NY: The Haworth Press, Inc. Also published as: *Cataloging & Classification Quarterly* 22(3/4).

Pearsall, Derek. 1985. "Editing medieval texts: Some developments and some problems." In Jerome J. McGann (Ed.), *Textual criticism and literary interpretation.* Chicago: University of Chicago Press, 92-106.

Popham, Michael. 1996. "Text encoding, analysis and retrieval." In Christine Mullings, Marilyn Deegan, Seamus Ross, and Stephanie Kenna (Eds.), 3-28.

Powell, Christina Kelleher and Nigel Kerr. 1997. "SGML creation and delivery: The Humanities Text Initiative," *D-Lib Magazine* July/August: <http://www.dlib.org/dlib/july97/humanities/07powell.html>.

Project Gutenberg. 1998. "Charles Dickens: Three ghost stories": <ftp://uiarchive.cso.uiuc.edu/pub/extext/gutenberg/extext98/3gsht/0.txt>.

Quigley, Ellie. 1995. *Perl by example.* Englewood Cliffs, NJ: Prentice-Hall.

Renear, Allen. 1992. "Representing text on the computer: Lessons for and from philosophy," *Bulletin of the John Rylands Library* 74(3):221-248.

Robinson, Peter. 1993a. *The digitization of primary textual sources.* Oxford: Office for Humanities Communication, Publication No. 4, Oxford University Computing Services.

Robinson, Peter. 1993b. "Redefining critical editions." In George Landow and Paul Delaney (Eds.), *The digital word.* Cambridge, MA: The MIT Press, 271-291.

Robinson, Peter. 1993c. "Manuscript politics." In Warren Chernaik, Caroline Davis, and Marilyn Deegan (Eds.), *The politics of the electronic text.* Oxford: Office for Humanities Communication, Publication No. 3, Oxford University Computing Services, 9-15.

Robinson, Peter. 1994. *The transcription of primary textual sources using SGML.* Oxford: Office for Humanities Communication, Publication No. 6, Oxford University Computing Services.

Robinson, Peter. 1996. "Image capture and analysis." In Christine Mullings, Marilyn Deegan, Seamus Ross, and Stephanie Kenna (Eds.), *New technologies for the humanities.* London: Bowker-Saur, 47-64.

Robinson, Peter and Elizabeth Solopova. 1993. "Guidelines for the transcription of the manuscripts of the Wife of Bath's Prologue." In Norman Blake and Peter Robinson (Eds.), *The Canterbury Tales Project occasional papers*, Volume 1. Oxford: Office for Humanities Communication, Oxford University Computing Services, 19-52.

Rockwell, Richard C. 1997. "The concept of the gateway library: A view from the periphery." In Lawrence Dowler (Ed.), *Gateways to knowledge: The role of academic libraries in teaching, learning, and research.* Cambridge, MA: The MIT Press, 109-122.

Rubinsky, Yuri and Murray Maloney. 1997. *SGML on the Web: Small steps beyond H.T.M.L.* Upper Saddle River, NJ: Prentice-Hall PTR.

St. Laurent, Simon. 1998. *XML: A primer.* Foster City, CA: MIS:Press.

Seaman, David. 1994. "Campus publishing in standardized electronic formats—HTML and TEI": <http://www.lib.virginia.edu/etext/articles/arl/dms-arl94.html>.

Seaman, David. 1996a. "Special collections digital image creation": <http://etext.lib.virginia.edu/helpsheets/specscan.html>.

Seaman, David. 1996b. "Selection, access, and control in a library of electronic texts." In Ling-Yuh Pattie and Bonnie Jean Cox (Eds.), *Electronic resources: Selection and bibliographic control.* Binghamton, NY: The Haworth Press, Inc., 75-84.

Shillingsburg, Peter L. 1993. "Polymorphic, polysemic, protean, reliable, electronic texts." In George Bornstein and Ralph Williams (Eds.), *Palimpsest: Editorial theory in the humanities.* Ann Arbor, MI: University of Michigan Press, 29-43.

Shillingsburg, Peter L. 1996. *Scholarly editing in the computer age*, Third edition. Ann Arbor, MI: University of Michigan Press.

Smith, Natalia and Helen R. Tibbo. 1996. "Libraries and the creation of electronic texts for the humanities," *College & Research Libraries* 57(6): 535-553.

Sperberg-McQueen, C.M. 1991. "Text in the electronic age: Textual study and text encoding, with examples from medieval texts," *Literary and Linguistic Computing* 6(1):34-46.

Sperberg-McQueen, C.M. 1994. "Bare bones TEI: A very very small subset of the TEI encoding scheme": <http://www.uic.edu/orgs/tei/intros/teiu6.html>.

Sperberg-McQueen, C.M. 1995. "The design of the TEI encoding scheme." In Nancy Ide and Jean Véronis (Eds.), *Text Encoding Initiative: Background and context.* Dordrecht, Netherlands: Kluwer, 17-39.

Sperberg-McQueen, C.M. and Lou Burnard (Eds.). 1994. *Guidelines for electronic text encoding and interchange.* Chicago: Text Encoding Initiative.

Starr, Susan S. 1998. "Building the collections of the California Digital Library," *Issues in Science and Technology Librarianship,* No. 17: <http://www.library.ucsb.edu/istl/98-winter/article2.html>.

Steinberg, S.H. 1974. *Five hundred years of printing,* Third edition. Revised by James Moran. Harmondsworth, Middlesex, UK: Penguin.

Tanselle, G. Thomas. 1983. "Classical, biblical, and medieval textual criticism and modern editing," *Studies in Bibliography* 36:21-68.

Task Force on Archiving of Digital Information. 1996. "Preserving digital information: Report of the Task Force on Archiving of Digital Information, commissioned by the Commission on Preservation and Access and the Research Libraries Group": <http://www.rlg.org/ArchTF/index.html>.

Tauber, James K. 1995. "Abandon all hope, ye who enter: A TEI novice recounts his experiences marking up *La Divina Commedia* and the Greek New Testament," *Text Technology* 5(3):225-233.

Travis, Brian E. and Dale C. Waldt. 1995. *The SGML implementation guide: A blueprint for SGML migration.* Berlin: Springer-Verlag.

Tuman, Myron. 1992. *Word perfect: Literacy in the computer age.* London: Falmer.

Turner, Eric G. 1977. *The typology of the early codex.* Philadelphia: University of Pennsylvania Press.

University of California. 1996. "The University of California Digital Library: A framework for planning and strategic initiatives": <http://sunsite.berkeley.edu/UCDL/summary.html>.

von Hagen, Bill. 1997. *SGML for dummies.* Foster City, CA: IDG Books.

Wainwright, Eric. 1996. "Digital libraries: Some implications for government and education from the Australian development experience": <http://www.nla.gov.au/nla/staff/paper/ew6.html>.

Wall, Larry. 1996. *Programming Perl,* Second edition. Bonn: O'Reilly.

Williamson, Hugh. 1983. *Methods of book design: The practice of an industrial craft,* Third edition. New Haven, CT: Yale University Press.

Wusteman, Judith. 1997. "Formats for the electronic library," *Ariadne* 8: <http://www.ukoln.ac.uk/ariadne/issue8/electronic-formats>.

Index

Order Your Own Copy of
This Important Book for Your Personal Library!

THE TEXT IN THE MACHINE
Electronic Texts in the Humanities

_____ in hardbound at $39.95 (ISBN: 0-7890-0424-0)

COST OF BOOKS _____

OUTSIDE USA/CANADA/
MEXICO: ADD 20% _____

POSTAGE & HANDLING _____
*(US: $3.00 for first book & $1.25
for each additional book)*
*Outside US: $4.75 for first book
& $1.75 for each additional book)*

SUBTOTAL _____

IN CANADA: ADD 7% GST _____

STATE TAX _____
*(NY, OH & MN residents, please
add appropriate local sales tax)*

FINAL TOTAL _____
*(If paying in Canadian funds,
convert using the current
exchange rate. UNESCO
coupons welcome.)*

☐ **BILL ME LATER:** ($5 service charge will be added)
(Bill-me option is good on US/Canada/Mexico orders only;
not good to jobbers, wholesalers, or subscription agencies.)

☐ Check here if billing address is different from
shipping address and attach purchase order and
billing address information.

Signature _____

☐ **PAYMENT ENCLOSED:** $_____

☐ **PLEASE CHARGE TO MY CREDIT CARD.**

☐ Visa   ☐ MasterCard   ☐ AmEx   ☐ Discover
☐ Diner's Club

Account # _____

Exp. Date _____

Signature _____

Prices in US dollars and subject to change without notice.

NAME _____

INSTITUTION _____

ADDRESS _____

CITY _____

STATE/ZIP _____

COUNTRY _____ COUNTY (NY residents only) _____

TEL _____ FAX _____

E-MAIL_____
May we use your e-mail address for confirmations and other types of information? ☐ Yes   ☐ No

Order From Your Local Bookstore or Directly From
The Haworth Press, Inc.
10 Alice Street, Binghamton, New York 13904-1580 • USA
TELEPHONE: 1-800-HAWORTH (1-800-429-6784) / Outside US/Canada: (607) 722-5857
FAX: 1-800-895-0582 / Outside US/Canada: (607) 772-6362
E-mail: getinfo@haworthpressinc.com
PLEASE PHOTOCOPY THIS FORM FOR YOUR PERSONAL USE.

BOF96